Microfauna Marina
Vol. 7

Akademie der Wissenschaften und der Literatur
Mathematisch-naturwissenschaftliche Klasse

Kommission für Zoologie

Akademie der Wissenschaften und der Literatur · Mainz · 1992

Microfauna Marina

Editor: Peter Ax

Vol. 7

Gustav Fischer Verlag · Stuttgart · Jena · New York · 1992

Gefördert mit Mitteln des
Bundesministeriums für Forschung und Technologie, Bonn,
und des Niedersächsischen Ministeriums für Wissenschaft
und Kunst, Hannover

Anschrift des Herausgebers
Prof. Dr. *Peter Ax,* II. Zoologisches Institut und Museum
der Universität Göttingen, Berliner Straße 28, D-3400 Göttingen

Abbildung auf dem Einband:
Hesionides bengalensis nov. spec. (Polychaeta)

Die Deutsche Bibliothek – CIP-Einheitsaufnahme

Microfauna marina / Akad. d. Wiss. u. d. Literatur,
Math.-Naturwiss. Klasse, Komm. für Zoologie;
Akad. d. Wiss. u. d. Literatur, Mainz. –
Stuttgart; Jena; New York: Fischer
ISSN 0176–3296
Erscheint unregelmäßig
Forts. von: Mikrofauna des Meeresbodens
Vol. I (1984) –

© 1992 by Akademie der Wissenschaften und der Literatur, Mainz
Gesamtherstellung: Röhm GmbH, Sindelfingen
Printed in Germany

ISBN 3-437-30698-7
US-ISBN 1-56081-359-8
ISSN 0176-3296

Inhalt/Contents

HANS-UWE DAHMS and MARK POTTEK: *Metahuntemannia* Smirnov, 1964 and *Talpina* gen. nov. (Copepoda, Harpacticoida) from the deep-sea of the high Antartic Weddell Sea with description of eight new species 7

WOLFGANG MIELKE: Description of some benthic Copepoda from Chile and a discussion on the relationships of *Paraschizopera* and *Schizopera* (Diosaccidae) 79

WOLFGANG MIELKE: Six representatives of the Tetragonicipitidae (Copepoda) from Costa Rica 101

WILFRIED WESTHEIDE: Neue interstitielle Polychaeten (Hesionidae, Dorvilleidae) aus dem Litoral des Golfs von Bengalen 147

PETER AX: *Promesostoma teshirogii* n. sp. (Plathelminthes, Rhabdocoela) from brackish water of Japan 159

PETER AX and ANDREAS SCHMIDT-RHAESA: The fastening of egg capsules of *Multipeniata* Nasonov, 1927 (Prolecithophora, Plathelminthes) on bivalves – an adaptation to living conditions in soft bottom 167

BEATE SOPOTT-EHLERS: *Coelogynopora sequana* nov. spec. (Proseriata, Plathelminthes) aus der Seine-Mündung 177

BEATE SOPOTT-EHLERS: *Coelogynopora faenofurca* nov. spec. (Proseriata, Plathelminthes) aus Wohnröhren des Polychaeten *Arenicola marina* 185

THOMAS BARTOLOMAEUS: Ultrastructure of the photoreceptors in certain larvae of the Annelida 191

THOMAS BARTOLOMAEUS: Ultrastructure of the photoreceptors in the larvae of *Lepidochiton cinereus* (Mollusca, Polyplacophora) and *Lacuna divaricata* (Mollusca, Gastropoda) 215

THOMAS BARTOLOMAEUS: On the ultrastructure of the cuticle, the epidermis and the gills of *Sternaspis scutata* (Annelida) 237

ULRICH EHLERS: Dermonephridia – modified epidermal cells with a probable excretory function in *Paratomella rubra* (Acoela, Plathelminthes) 253

ULRICH EHLERS: On the fine structure of *Paratomella rubra* Rieger & Ott (Acoela) and the position of the taxon *Paratomella* Dörjes in a phylogenetic system of the Acoelomorpha (Plathelminthes).................... 265

ULRICH EHLERS: "Pulsatile bodies" in *Anaperus tvaerminnensis* (Luther, 1912) (Acoela, Plathelminthes) are degenerating epidermal cells 295

ULRICH EHLERS: No mitosis of differentiated epidermal cells in the Plathelminthes: mitosis of intraepidermal stem cells in *Rhynchoscolex simplex* Leidy, 1851 (Catenulida) 311

CORNELIA WENZEL, ULRICH EHLERS and ALBERTO LANFRANCHI: The larval protonephridium of *Stylochus mediterraneus* Galleni (Polycladida, Plathelminthes): an ultrastructural analysis 323

Short Communication

PETER AX: Plathelminthes from brackish water of Northern Japan: no identical species with the corresponding boreal community 341

Metahuntemannia Smirnov, 1946 and *Talpina* gen. nov. (Copepoda, Harpacticoida) from the deep-sea of the high Antarctic Weddell Sea with a description of eight new species.

Hans-Uwe Dahms and Mark Pottek

Contents

Abstract	7
A. Introduction	8
B. Material and methods	9
C. Description of species	10
Metahuntemannia texturata sp. n.	10
Metahuntemannia beckeri sp. n.	18
Metahuntemannia spinipes sp. n.	24
Talpina gen. n.	28
Talpina noodti sp. n.	29
Talpina fodens sp. n.	35
Talpina furcispina sp. n.	40
Talpina bathyalis sp. n.	47
Talpina pectinata sp. n.	52
D. Discussion	62
Zusammenfassung	76
References	77

Abstract

The taxon *Metahuntemannia* Smirnov, 1946 is subdivided into the two sister-taxa *Metahuntemannia* s. str. and *Talpina* gen. n. The findings of new constitutive characters and a reevaluation of characters already known lead to the erection of the new taxon *Talpina* gen. n. which represents the former *talpa*-group *sensu* BECKER (1979) of *Metahuntemannia* s. l. Representatives of the *spinosa*-group which have been described earlier are belonging to *Metahuntemannia*. *Beckeria* Por, 1986 is synonymized with *Metahuntemannia*. Material obtained during 3 expeditions of RV Polarstern to the Weddell Sea yielded 8 new species revealing a considerably high diversity. Three of the 8 new species are belonging to *Metahuntemannia*, namely *M. texturata, M. beckeri*, and *M. spinipes*; five of the new species are belonging to *Talpina*, namely *T. noodti, T. fodens, T. furcispina, T. bathyalis*, and *T. pectinata*. All the species must be well adapted for an inbenthic mode of life as it is evidenced by a cylindrical body shape, stout appendages, spiniform setae on the antennules, reductions in the locomotory leg endopodites and setal lengths of the exopodites. In *Talpina* it is especially the P 1 which is specialized for a digging function as indicated by an elongated coxa, a medially flexed exopodite and stout exopodal segments armed with spiniform setae forming a shovel together with the reduced endopodite. Obviously, females of both taxa do not carry egg-sacs. Instead, they probably deposit their eggs soon after laying them. The sex ratio is highly female biassed as in other copepods of the deep sea. A reevaluation of zoogeography and phylogenetic relationships among the representatives of *Metahuntemannia* comprising now of 17 species and *Talpina* comprising of 11 species as well as a key for the identification is given.

A. Introduction

In the course of a study of Antarctic meiobenthos during three expeditions of RV Polarstern, meiofauna was collected in the eastern Weddell Sea of the high Antarctic. The particular expeditions were as follows: Ant V/3 (29. 9. to 14. 12. 1986), Ant VII/4 (13. 1. to 10. 3. 1989) and Ant VIII/6 (13. 3. to 1. 5. 1990). The harpacticoid assemblages of the high Antarctic are dominated by Cletodidae s. l., Diosaccidae, Tachidiidimorpha and Tisbidimorpha. One of the most peculiar taxa of Cletodidae is *Metahuntemannia* Smirnov, 1946, e. g. for its deep-sea occurrence, sparse distribution, remarkable sexual dimorphism and its first locomotor thoracopod being transformed to a burrowing device. Careful inspection and the findings of new characters of the two groups – the *spinosa*-group and the *talpa*-group according to BECKER[1] and BECKER (1979) – led to the establishment of a new taxon *Talpina* gen. n. comprising of those species formerly belonging to the *talpa*-group. In the course of the investigation three new species of *Metahuntemannia* Smirnov, 1946 and five new species of *Talpina* were found and described in the present investigation.

---

[1] BECKER, K. H. (unpubl. Ph. D. thesis). Eidonomie und Taxonomie abyssaler Harpacticoidea (Crustacea, Copepoda). Dissertation Univ. Kiel, 1972: 1–163.

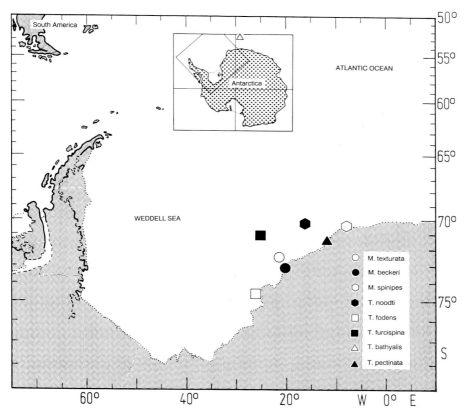

Fig. 1. Map of stations in the eastern Weddell Sea (Antarctica) where species of *Metahuntemannia* s. str. and *Talpina* gen. n. have been found.

B. Material and methods

During three cruises of RV Polarstern meiofauna has been collected from sea-bottom samples using Agassiz-trawl (AGT), box corer (GKG), multi box corer (MG) and dredge (D), namely during the German Ant V/3, Ant VIII/6 and European Ant VII/4 (Epos leg 3) expeditions. Specimens of *Metahuntemannia* and *Talpina* were found at the locations mentioned in the species descriptions and these are shown in Fig. 1. Collections were made by stirring up substrate in a beaker and decanting it over a screen (80 μm mesh size). Specimens were preserved in 4 % formaldehyde and transferred into W 15 embedding medium (C. Zeiss) for dissection and identification. The type specimens are within the collection of the "Zoologisches Museum der Universität Hamburg" under the catalogue number given under 'material'! Before dissection the habitus was drawn from whole mounts and body size measurements were made. Length is given including rostrum and caudal rami and without rostrum (to base of rostrum) but including caudal rami (in brackets). All figures were drawn with the aid of a camera lucida. The terms pars incisiva, pars molaris and lacinia mobilis are omitted in the description of the mandibular precoxa (according to MIELKE 1984). Abbreviations used in the text are: Ro = rostrum; A 1 = antennula; A 2 = antenna; Md = mandible; Mx 1 = maxillula; Mx 2 = maxilla; Mxp = maxilliped; P 1 – P 6 = legs 1 – 6; seg. III = 3rd segment; I–III = segments I to III; enp, exp = endopodite, exopodite; enp 2 = 2nd segment of endopodite; benp = baseoendopodite; Cu.r. = caudal ramus; gen.f. = genital field; C I – C VI = first to adult copepodid; f = female; m = male. Formula used for the armature of the antennula, e. g.: 0.–3 + Ae.– ... = 1st segment with no seta/2nd segment with 3 setae and 1 aesthetasc.

C. Description of species

Metahuntemannia texturata sp. n.

(Figs. 2–6)

Material. One female collected during Ant V/3 by AGT 10/513 (18. 10. 86) from 250 m depth (K–35094 *a+ b*).

Locality. 72° 38,3' S – 20° 25,6' W

Female
Body length 875 (840) µm and maximum body width 220 µm.

Body (Fig. 2A) fusiform with large cephalosome. Caudal margins of urosomal somites except ultimate one encircled with a double row of spinules. Last thoracal somite only laterally and dorsally with double spinular row. Penultimate segment of thorax with a single dorsal row of spinules. Denticles on cuticular surface of the body somites not clearly arranged in distinct rows but in a diffuse pattern. Anal somite small and distally with lateral spinular patches. Genital somite undivided.

Rostrum (Fig. 4) not plough-like but of almost triangular shape and dorsally ornamented with dot-like naps. Paired seta inserting anteriorly.

Caudal rami (Figs. 6 A, B) subterminally widened by a huge inner ventrolateral lobe. Edge of lobe fringed with about ten spinules. Rami with six setae. Dorsal one plumose and biarticulated at base. Both the lateral ones inserting on

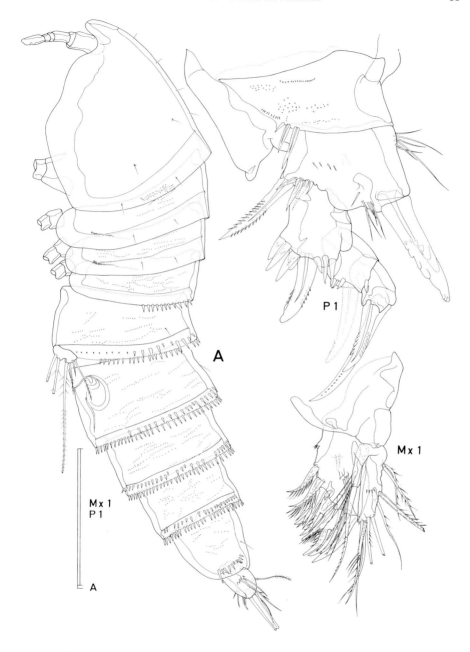

Fig. 2. *Metahuntemannia texturata* sp. n. Female. Habitus in left lateral view (A); right P 1 in anterior view; maxillula. Scale bars for appendages and habitus 50 and 200 μm respectively.

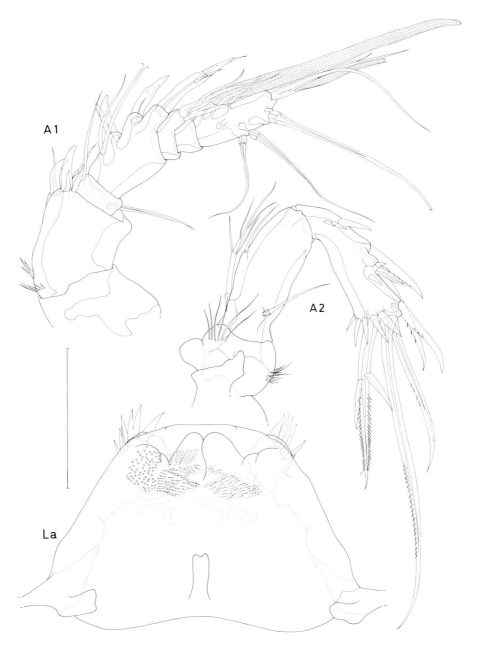

Fig. 3. *Metahuntemannia texturata* sp. n. Female. Antennula; antenna; labrum in anterior view. Scale bar 50 μm.

a small outer dorsal protuberance. Terminal major seta stout, proximally with furrows along longitudinal axis. This seta accompanied by a slender outer one and an inner spine-like seta. Distal edge of rami ventrally with four large spinules. Cuticular naps not only on surface of the rami but also on the major, stout, terminal seta following the furrows.

Antennula (Figs. 3, 4) six-segmented. All the setae are bare and comparatively short. Segment I bearing two rows of spinules proximally on inner margin and four armature elements at distal inner corner consisting of one seta and three spines. Dorsal surface ornamented with naps. Segment II small with two bare setae at inner and a single one at outer distal corner. Four thickened spinules on dorsal surface. Segment III with two setae proximally and three stout denticulated spines inserting on inner protuberances. Distalmost spine largest furnished with 4 larger denticles. Segment IV small with the distal inner process carrying a long aesthetasc and a short seta, both with common base. Segment V small with a single inner seta. Segment VI bearing three bare setae on inner margin. Outer margin with three articulated setae. Distal edge with slender aesthetasc chitinized at base and three bare setae, two of which share a common base with the aesthetasc.

Armature formula: 4.–3.–5.–1+Ae.–1.–9+Ae.

Antenna (Fig. 3). Coxa ornamented with two patches of spinules. Allobasis proximally with geniculate seta bearing four long spinules. Inner margin additionally with a single bare seta. Exopodite small and 1-segmented with a single bare seta. Endopodite with four non-geniculate setae on distal edge, all more or less pinnate, the outermost additionally with two large spinules, longer than allobasis and endopodite together, innermost claw shortest. Three strong spines on inner margin of endopodite. Three patches of large spinules, two along inner margin, the other along distal edge.

Labrum (Fig. 3) trapezoid with naps on anterior surface. Posterior surface with symmetrical spinular ornamentation. Both distal corners carrying a row of spinules each. Distal margin unornamented.

Mandible (Fig. 4). *Corpus mandibulae* with comb of spinules proximally. Cutting edge with two stout outer bidentate spines. Proximal edge bearing two slender bifurcate spines and one pinnate spine-like seta. Palp small and sub-cylindrical, comprising of basis and 1-segmented endopodite. Basis unarmed. Endopodite with two terminal setae, the outer longer one pinnate along middle third. Exopodite absent.

Maxillula (Fig. 2). Precoxal arthrite broad with eight claw-like spines, most of them pinnate or bidentate. Anterior surface furnished with two slender setae. Endite of coxa small with terminal claw-like curved seta and a subterminal set of long setules. Basis carrying five setae at terminal margin, two of which inserting on small outer protuberances. Endopodite and exopodite represented each by two bipinnate setae.

Maxilla (Fig. 4). Syncoxa proximally with inner and outer lateral rows of spinules. Two subcylindrical endites. The proximal one with a subterminal setule fan and a set of four terminal armature elements containing two short setae, a serrate spine, and another spine bearing three long spinules at tip. Distal endite with five slender setae terminally. Basis terminally with stout serrate claw and two setae. Endopodite consisting of a small segment with four bare setae.

Maxilliped (Fig. 4) prehensile. Basis with inner lateral row of spinules and distal inner rod-like spine bearing four longer spinules at tip. Endopodite 2-segmented; first segment with long spinules along inner margin and a row of

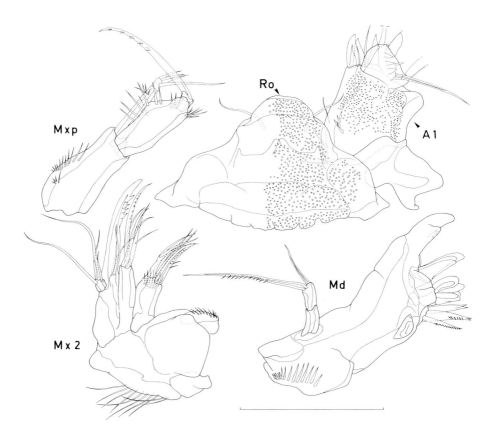

Fig. 4. *Metahuntemannia texturata* sp. n. Female. Maxilliped; rostrum and proximal portion of antennula; maxilla; mandible. Scale bar 50 μm.

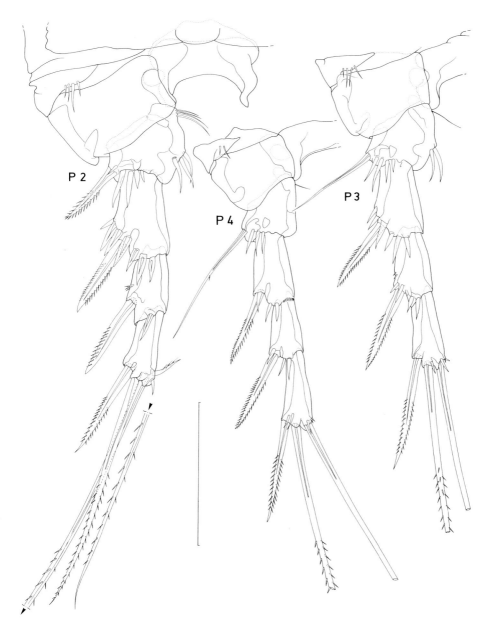

Fig. 5. *Metahuntemannia texturata* sp. n. Female. Right thoracic locomotor appendages P 2–4. Scale bar 50 μm.

smaller spinules subdistally. Second segment with long claw-like seta bipinnate along distal half. Claw longer than first segment.

P 1 (Fig. 2). Coxae of both counterparts not fused bearing several denticular ornamentations and three large spinules at the outer distal corner. Basis with a stout bipinnate spiniform seta at the outer distal corner and a row of large spinules above insertion plane of exopodite. Sets of setules between insertions of both rami and proximally along the inner margin. Endopodite represented by a large blunt spine (bs) with two rows of lobular emarginations. Exopodite three-segmented and slightly bent inwards. First segment the largest one carrying a stout blunt spiniform seta and several spinules on outer margin. Second segment with a similar spine and smaller spinules. Third segment the smallest one with only two stout spines the terminal longest.

P 2–4 (Fig. 5) uniramous. Intercoxal plates arched. Precoxae as small triangular scales with up to four long spinules. Coxae large, almost square in shape, and unornamented. Bases with outer seta developed into a bipinnate spine of P 2, almost bare at P 3 and P 4. Row of strong spinules above insertion planes of the exopodites. Endopodite possibly represented by long spinules of P 2 and P 3. P 4 without any trace of endopodite. Exopodite three-segmented, first segment is the largest one. Terminal setae long and bipinnate. Spine and seta formulae as follows:

	Exopodite	Endopodite
P 1	0.0.020	bs
P 2	0.0.021	–
P 3	0.0.021	–
P 4	0.0.021	–

P 5 (Fig. 6) with small bean-shaped intercoxal plate. Baseoendopodite and exopodite fused to a bilobed plate. Inner lobe with two setae refering to original endopodite. Outer lobe with seta on distal outer protuberance representing the outer seta of the baseoendopodite and three setae of different length representing the exopodite. Innermost of these setae is by far the longest.

Genital field as in Fig. 6.

Etymology. The species name refers to the morphological characteristic of its cuticle being napped and is derived from the Latin *textum*, meaning tissue.

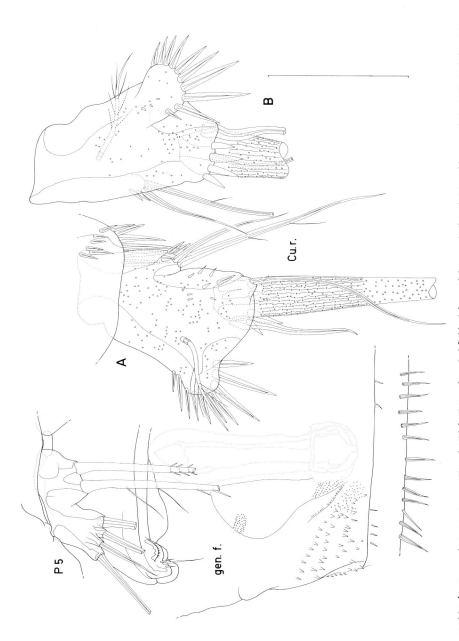

Fig. 6. *Metahuntemannia texturata* sp. n. Female. Right P 5 and genital field; right caudal ramus in dorsal (A) and in ventral view (B). Scale bar 50 μm.

Metahuntemannia beckeri sp. n.

(Figs. 7–10)

Material. One female collected during leg Ant V/3 by AGT 10/573 (6.11.86) from 390 m depth (K–35095).
Locality. 73° 06,8' S – 20° 15,0' W

Female

Body length 825 (805) µm and maximum body width 270 µm.

Body (Fig. 7) fusiform and similar to that of *M. texturata* but posterior margin of urosomal somites except the ultimate one ornamented only by a single spinular row dorsally and laterally. Last thoracic somite only dorsally fringed with spinules. Anal somite different from *M. texturata* in the more intense spinular trimmings. Anal operculum furnished with a distal row of small spinules.

Rostrum similar to that of *M. texturata* but without cuticular naps.

Caudal rami (Figs. 10 A, B) extremely chitinized and carrying six setae. Dorsal one plumose and biarticulated at base. Both, the lateral ones bare and of different length, attached to an outer projection. From the two other terminal setae the inner one is longest. Spinular row dorsally above insertion of terminal setae. Inner surface of rami distally extended into a ventrolateral lobe fringed with long spinules. Distal outer corner of rami with a few long spinules ventrally.

Antennula (Fig. 8). Six-segmented and very similar to that of *M. texturata*. Ventral surface of segment I with patches of little nobs. Segment II small with three bare setae at distal inner corner. Three stout spines of segment III with even heavier denticular armature than in *M. texturata*. Aesthetasc of segment IV chitinized at base and supported by a cuticular ledge. Armature of the two distal segments identical to that in *M. texturata*.

Armature formula: 4.–3.–5.–1 + Ae.–1.–9 + Ae.

Antenna (Fig. 8) almost identical to that of *M. texturata*. Exopodite minute with a single plumose seta.

Mandible (Fig. 8). Similar to that of *M. texturata*. Cutting edge distally with a stout tooth. Proximal edge with two slender bifurcate spines; a seta is missing. Palp two-segmented, basis longer than endopodite, the latter with two plumose setae terminally. Exopodite absent.

Maxillula (Fig. 8). Syncoxa richly ornamented. Precoxal arthrite terminally with eight claw-like spines, all except the two outer ones furnished with a spinule row. Two slender bare setae on posterior surface. Coxal arthrite with two terminal setae. Basis carrying two geniculate setae and a pinnate spine on distal margin. Two setae on subdistal projection. Endopodite and exopodite each represented by two spinulose setae on small lobular protuberances.

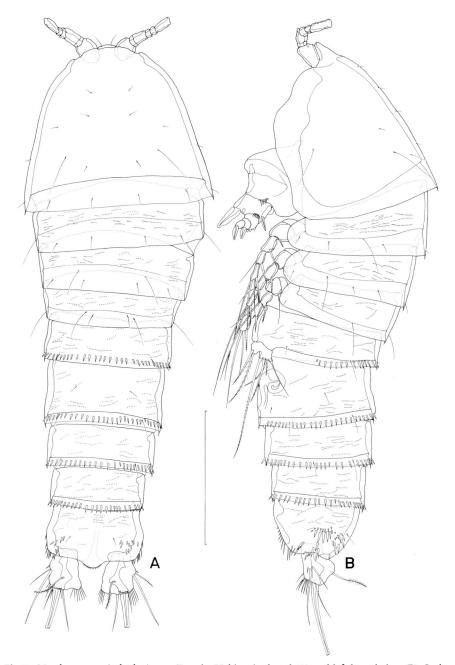

Fig. 7. *Metahuntemannia beckeri* sp. n. Female. Habitus in dorsal (A) and left lateral view (B). Scale bar 200 μm.

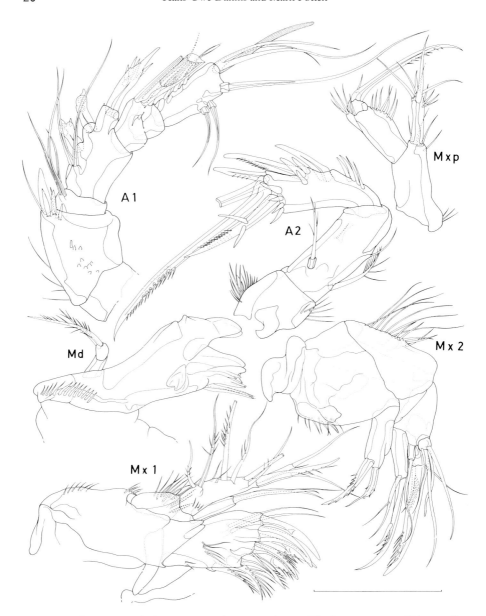

Fig. 8. *Metahuntemannia beckeri* sp. n. Female. Antennula; maxilliped; antenna; mandible; maxilla; maxillula. Scale bar 50 μm.

Maxilla (Fig. 8). Syncoxal ornamentation as in *M. texturata*. Two endites, proximal one with two bipinnate spines, distal endite smaller with three curved armature elements. Basis with one stout terminal claw subdistally pinnate. Two setae at base of claw, one of them is plumose. Third seta proximally on small cylindrical process. Endopodite with a small segment bearing four bare setae of different lengths.

Maxilliped (Fig. 8) prehensile. Distal inner spine of basis rodlike and distally furnished with spinules of varying lengths. First segment of endopodite with a spinule row along inner margin and outer fan of long spinules at base of second segment. This segment with pinnate terminal claw-like seta and a spinule on subdistal projection. Claw longer than endopodite.

P 1 (Fig. 10). Coxae of both counterparts not fused, carrying an oblique comb of spinules near the outer distal corner. Basis with stout outer spine. Spinule row above insertion of endopodite and exopodite respectively. Patch of spinules between insertions of both rami. Pore with bag-like tube on anterior surface. Endopodite represented by a single blunt spine. Exopodite three-segmented and slightly bent inwardly. First segment largest one, with a stout blunt spine and two sets of differently shaped smaller spinules along outer margin. Second segment with a stout outer spine and two small bifurcate spinules distally. Third segment with two terminal stout spines. All spine-like setae unornamented.

P 2–P 4 (Fig. 9). P 2 with a small 2-segmented endopodite. P 3 and P 4 uniramous. Coxae large with outer lobe and oblique spinule row on outer margin in P 2 and P 3. Pore with bag-like tube on anterior surface in P 2. Bases with a spinule row above insertion of exopodite, in P 2 also above insertion of endopodite. Outer seta modified into a short bipinnate spine in P 2; this seta lacking in P 4 (as in *T. furcispina*). Endopodite of P 2 with a small proximal segment and a minute distal one bearing a single setule. Endopodite of P 3 possibly represented by a protuberance with long spinules, that of P 4 by an unarmed lobe of the basis. Exopodite three-segmented. Spine and seta formulae as follows:

	Exopodite	Endopodite
P 1	0.0.020	bs
P 2	0.1.022	0.010
P 3	0.1.022	–
P 4	0.1.022	–

P 5 (Fig. 10) with baseoendopodite and exopodite fused to a bilobed plate. Inner lobe with two setae representing the endopodite. Exopodite is referred to by four bipinnate setae on outer lobe. Innermost seta is far the longest. Outer seta of baseoendopodite on cylindrical process.

Genital field as depicted in Fig. 10.

Etymology. The species is named after the late Dr. K. H. Becker who described six new species of *Metahuntemannia*.

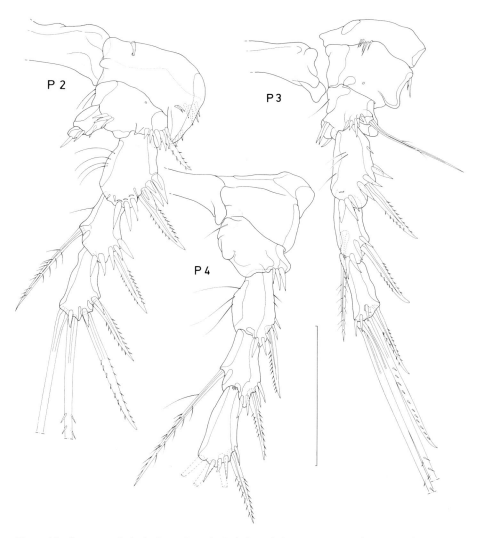

Fig. 9. *Metahuntemannia beckeri* sp. n. Female. Left thoracic locomotor appendages P 2–4 in anterior view. Scale bar 50 μm.

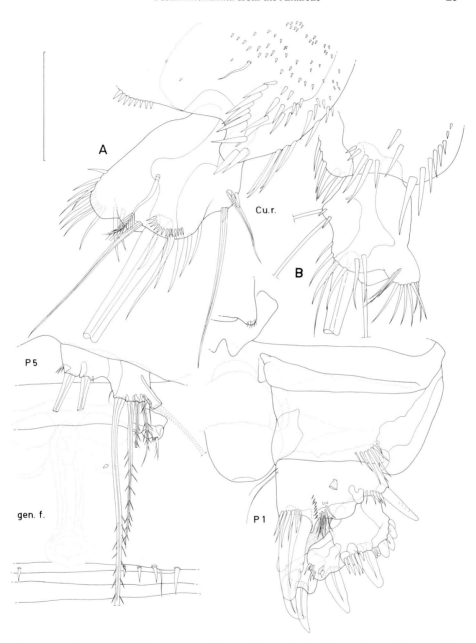

Fig. 10. *Metahuntemannia beckeri* sp. n. Female. Caudal ramus in dorsal (A) and ventral view (B); left P 5 and genital field; left P 1 in anterior view. Scale bar 50 μm.

Metahuntemannia spinipes sp. n.

(Figs. 11–13)

Material. One female collected during Ant V/3 by D 10/504 (12. 12. 86) from 270 m depth (K – 35096).
Locality. 70° 30,40' S – 08° 02,50' W

Female
Body length 740 (720) µm and maximum body width 250 µm.
Body fusiform and very similar to that of *M. beckeri*. Anal operculum distally fringed with five spinules (Fig. 13). Rostrum (Fig. 11) small and triangular; distal edge with two bare setae as in all other representatives of *Metahuntemannia*.
Caudal rami (Fig. 13 A, B) square with five setae. Dorsal seta bare and biarticulated at base. Lateral setae bare inserting on an outer projection, the more distal one almost three times as long as the other one. The major terminal seta is accompanied by a small inner seta inserting from a small process. Outer terminal seta reduced to a minute spine. Dorsolateral lobe close to biarticulated seta fringed with some spinules.
Antennula (Fig. 11) six-segmented. Segment I bearing four spines and a seta at distal inner corner. Three rows of spinules on inner margin. Segment II small with three setae at distal inner corner. Segment III with two setae and three spines along inner margin, the distal two inserting from a process, the distalmost one with 2 spinules midlength. Its armature not as intense as in *M. texturata* and *M. beckeri*. Segment IV short with a long aesthetasc supported proximally by a cuticular ledge and sharing a common base with its accompanying long seta. Anterior surface with another seta. Segments V and VI as in *M. texturata* and *M. beckeri* but last segment with additional tiny spinule (probably remnant of the fourth seta on the outer margin of the terminal segment).
Armature formula: 5.–3.–5.–2 + Ae.–1.–9(10) + Ae.
Antenna (Fig. 11). Allobasis with two plumose setae on inner margin. Exopodite present as a small segment with one plumose seta. Second segment of the endopodite resembling that of *M. texturata* and *M. beckeri* but outermost spiniform seta on distal edge not ornamented with spinules.
Mandible (Fig. 11). Cutting edge with four teeth, the two distal ones tridentate, the others bidentate, followed by a set of spinules and four spines. Proximal edge with a slender bifurcate spine and a short pinnate seta. Inner protuberance well developed; palp lost during preparation.
Maxilla (Fig. 11). Ornamentation of syncoxa similar to that of *M. texturata* and *M. beckeri*. Two cylindrical endites, the distal one with three terminal setae accompanying a slender claw armed with some spinules. Basis with terminal claw

and three setae proximally. Endopodite as a small square segment bearing four bare setae.

Maxilliped (Fig. 11). Basis on posterior surface with two spinule rows, one of which next to spine at distal inner corner. This spiniform, remarkably large and rod-like seta distally furnished with spinules of varying lengths. First endopodite segment richly ornamented with long spinules. Second segment of endopodite small, and terminal seta mostly reduced, not claw-like. This maxilliped can hardly be used in a prehensile function.

P 1 (Fig. 11). Both counterparts not fused. Coxa and basis as in *M. beckeri* but outer spine of basis bipinnate. Endopodite represented by a stout blunt spine. Exopodite three-segmented. First segment is largest bearing one blunt spine and

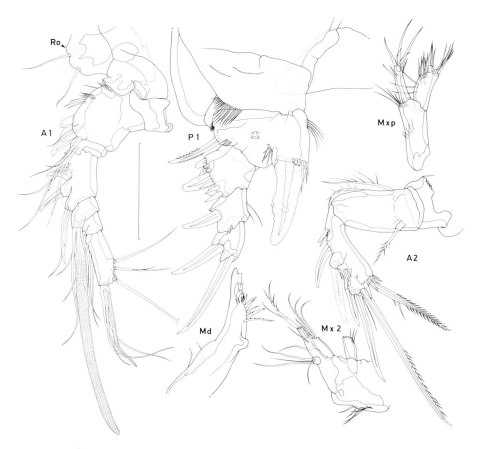

Fig. 11. *Metahuntemannia spinipes* sp. n. Female. Antennula and rostrum in ventral view; left P 1 in posterior view; maxilliped; antenna; mandible; maxilla. Scale bar 50 μm.

some smaller spinules along outer margin. Second segment with outer spine and a few distal spinules. Distal segment with two terminal longer spines.

P 2–P 4 (Fig. 12) biramous. Intercoxal plates yoke-shaped. Triangular precoxae fringed with long spinules at distal edge. Coxae rectangular with two spinule sets at outer margins but unornamented in P 4. Basis of P 2 with one long outer seta modified into a bipinnate spine. A row of long spinules above insertion of endopodite and exopodite respectively. Pore with bag-like tube on anterior surface. Endopodite two-segmented, second segment slender and slightly longer than first one. Both segments richly ornamented with long spinules. Exopodite three-segmented, first segment the largest one. Especially in P 2 and P 3 all segments bearing thickened spinules at their distal outer corners. Setae of both rami plumose. Outer terminal seta of exopodites fringed with an

Fig. 12. *Metahuntemannia spinipes* sp. n. Female. Right thoracic locomotor appendages P 2–4 in anterior view. Scale bar 50 µm.

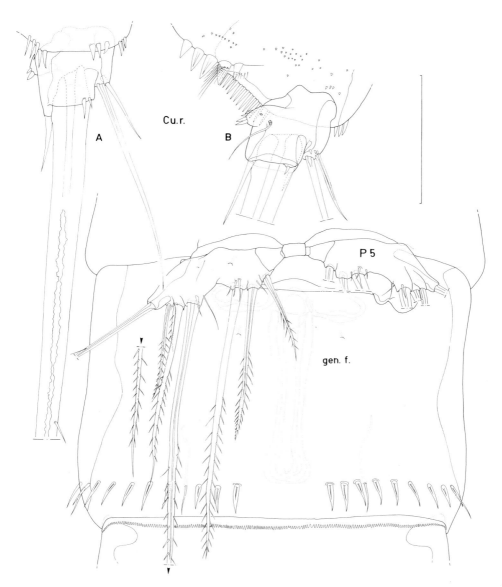

Fig. 13. *Metahuntemannia spinipes* sp. n. Female. Caudal ramus in ventral (A) and dorsal view (B); P 5 and genital field in ventral view. Scale bar 50 μm.

inner row of long spinules and on opposite side with short spinules. Spine and seta formulae as follows:

	Exopodite	Endopodite
P 1	0.0.020	bs
P 2	0.1.122	0.220
P 3	0.1.222	0.220
P 4	0.1.222	0.120

P 5 (Fig. 13) with rectangular intercoxal plate. Baseoendopodite and exopodite fused to a bilobed plate, each lobe with a pore. Inner lobe carrying three bipinnate setae increasing in length from inside to outside and refering to the endopodite. Outer lobe with four setae on outer distal process. Exopodite represented by three bipinnate setae of outer lobe increasing in length from outside to inside. Outer seta of baseoendopodite arising from a protuberance.

Genital field as in Fig. 13.

Etymology. The species name refers to the morphological characteristic especially of the swimming legs being richly spinulated and is derived from the Latin *spina* and *pes*, meaning thorn and foot respectively.

Talpina gen. n.

Consisting of all representatives belonging to the *talpa*-group *sensu* BECKER (1979) of *Metahuntemannia* Smirnov, 1946.

Generic diagnosis

Habitus vermiform; somites with a peculiar pattern of distinct and symmetrical denticle rows on surface; anal somite elongate. Rostrum plough-like. Segment III of antennule not ornamented with denticulated spiniform setae. Basis of mandible bearing 1 seta; its endopodite with 1–3 setae; proximal edge of *corpus mandibulae* with 2 stout and curved elements. Basis of maxilliped with a long spine, spinulated along the tapering tip. Coxae of P 1 fused; endopodite present (as a 1-segmented ramus with 1 seta in the females and 2-segmented in the males); exopodite in the females either 1- or 3-segmented, strongly bent inwardly. P 5 either reduced to a single lobe only bearing the outer seta of the baseoendopodite or with a separate exopodite. Caudal rami small, longer than wide, with a small unornamented dorsal lobe next to dorsal seta; pore with bag-like tube on dorsal surface.

Etymology. The generic name is derived from the Latin *talpa*, meaning mole. Gender: feminine.

Genotype. *Talpina curticauda* (Becker, 1979).

Talpina noodti sp. n.

(Figs. 14–17)

Material. One female collected during Ant V/3 by AGT 10/627 (26. 11. 86) from 430 m depth (K – 35097 $a + b$).

Locality. 72° 03,5' S – 15° 27,6'W

Female

Body length 1530 (1510) μm and maximum body width 300 μm. Urosome length 80 μm of which last somite covers more than a third.

Body (Fig. 14) vermiform. All somites except cephalosome with a complex pattern of spicule rows. Spiculation especially pronounced at urosome. Posterior margin of all somites except the last two ones of urosome carrying two to four pairs of spinules.

Penultimate somite of urosome without spinules and last one with a paired dorsolateral seta at the end of the first third of the somite. Genital somite undivided, bearing a paired comb of spinules ventrally at the posterior margin as in the next somite. Last somite is the largest one of the urosome bearing three peculiar ventrolateral spinule fields of circular shape on both sides.

Rostrum (Fig. 14) prominent, bifid, and of plough-like shape.

Caudal rami (Fig. 15) small, two times longer than broad, carrying two slender plumose setae of equal length midlength on the outer margin. Dorsal seta plumose and biarticulated, half as long as the two setae of outer margin, inserting close to a small inner dorsal lobe. Terminal seta accompanied by a short outer bare seta with thickened base and a small inner spine. Two pores next to dorsal seta, at least one of them with a bag-like widened tube.

Antennula (Fig. 15) short, six-segmented. Segment I with several spinules at inner margin and spinulose, spine-like seta at distal inner corner. Segment II armed with two spiniform setae, one of which spinulose, and two short and crooked spines at distal edge. Segment III bearing five elements at inner margin: one small spine, a slender bare seta, and three spine-like bare setae, the distal-most inserting on a short process. Segment IV terminating into a process at distal inner corner bearing a long aesthetasc supported by a cuticular ledge. Aesthetasc with a subdistal setuliform outgrowth at tip; a long bare seta and a small crooked spine accompanying the aesthetasc. Segment V small, with a spinelike bare seta at distal inner corner. Segment VI long, with two spines and a spine-like bare seta at inner margin, two bare setae articulated at base and two distal curved spines at outer margin, and a terminal small aesthetasc with chitinized base accompanied by three bare setae.

Armature formula: 1.–6.–5.–2 + Ae.–1.–10 + Ae.

Mandible (Fig. 16). Cutting edge with a tuft of long spinules and five spines.

Of adjacent two spines one also is bidentate, the other slender with two distal spinules. Proximal edge bearing two elements with one-sided spinulation and one smaller bare seta proximally. Palp two-segmented. Basis with pinnate seta at inner corner and a slender bare setule. Endopodite with three plumose setae, the

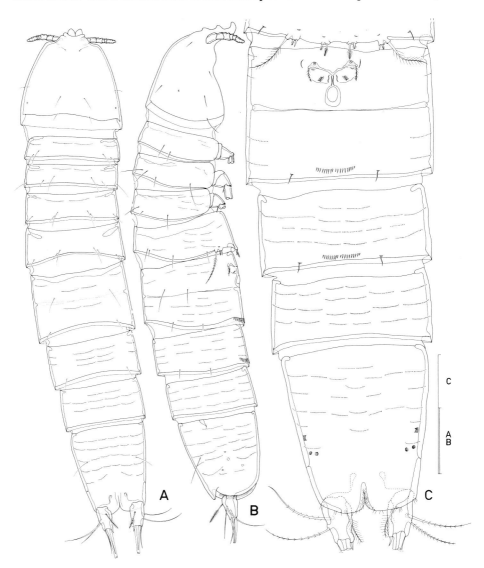

Fig. 14. *Talpina noodti* sp. n. Female. Habitus in dorsal (A) and lateral view (B); urosome in ventral view (C). Scale bars 50 μm.

terminal one almost three times as long as the other two. Innermost seta inserting from a short protuberance. Exopodite absent.

Maxillula (Fig. 16 A, B). Arthrite of precoxa with eight spiniform elements, all except the third to outermost plumose (Fig. 16 A). Arthrite of coxa lost during preparation. Basis with three terminal bare setae (Fig. 16 B). Endopodite represented by two bare setae, the distalmost reduced in size. Exopodite appearing as two bare setae and one smaller spinulose seta.

Maxilla (Fig. 15). Syncoxa ornamented with a spinule row on outer margin and a semicircular arrangement of spinules on the inner margin. Two endites,

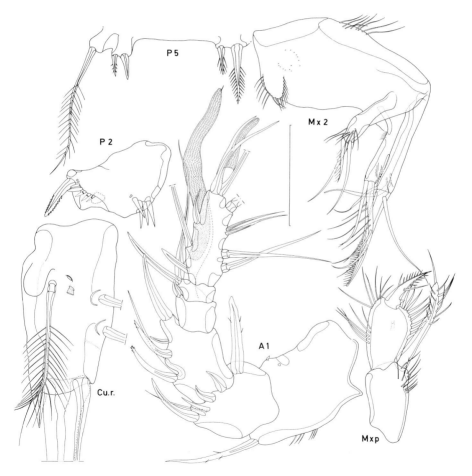

Fig. 15. *Talpina noodti* sp. n. Female. Right P 5; maxilla; left P 2 basis; right caudal ramus; antennula; maxilliped. Scale bar 50 μm.

each bearing a spinulose seta and two bare setae. Proximal endite furnished with a subdistal row of spinules. Basis bearing an endite with two bare setae and a claw-like terminal seta carrying one single row of spinules. Endopodite represented by a single blunt seta and two minute setules.

Maxilliped (Fig. 15) prehensile. Basis bearing a long plumose spine surrounded at base by a patch of spinules at the inner distal corner. First segment of endopodite fringed with long spinules all along the inner and outer margin and carrying a terminal claw-like seta probably fused with the second segment of the endopodite. Claw bipinnate along apical half; pinnation beginning with a longer

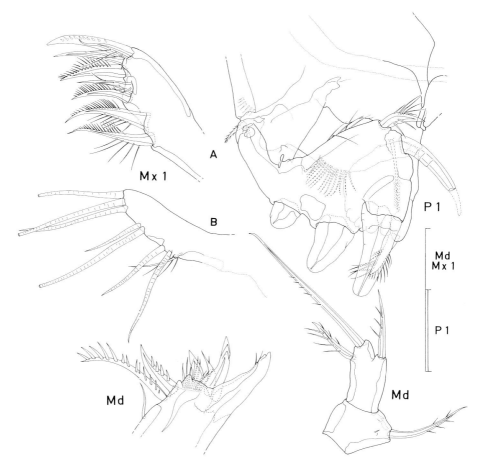

Fig. 16. *Talpina noodti* sp. n. Female. Precoxal arthrite (A) and basis (B) of maxillula; P 1; mandibular cutting edge and mandibular palp. Scale bar 50 µm.

spinule on each side. Base of claw widened, carrying a small spine on inner surface, probably representing the former 2nd segment of the endopodite.

P 1 (Fig. 16) with striking shape-modifications of original biramous swimming leg. Coxae of both counterparts completely fused. Basis with fan of long spinules between insertions of rami and plumose seta accompanied by some spinules on posterior surface above insertion of endopodite; outer margin with a stout plumose seta. Exopodite one-segmented, inwardly bent and distally widened. Three strong blunt spines along outer margin with a terminal pore increasing in size from proximal to distal position. Inner margin furnished with a spinule row and two spines, the proximalmost smaller and plumose. Endopodite as a broad one-segmented plate bearing on terminal spinulated seta. Subterminal outer edge showing flute-like structures.

P 2–P 4 (Fig. 17) uniramous. Intercoxal plates arched, attached to the whole inner edge of coxae. Precoxae developed as a scale of triangular shape with the most acute angle pointing medially. Base line of triangle fringed along the middle third with up to 17 tiny spinules. Coxae with an oblique spinular row on outer edge of posterior surface. Bases of about equal size as coxae carrying a long plumose seta on outer margin, which is modified in P 2 to a short bipinnate spine (Fig. 15). Oblique spinular row on anterior surface above insertion of exopodites. Endopodites possibly represented by a row of long spinules at inner lateral edge of bases. Exopodite three-segmented, spine and seta formulae as follows:

	Exopodite	Endopodite
P 2	0.1.221	–
P 3	0.1.222	–
P 4	0.1.222	–

P 5 (Figs. 14 C, 15) consisting of two lobes at posterior margin of last thoracal segment each carrying a set of setae. Each lobe bears a pore at base. Outer lobe with three setae, the longest outer one is a plumose seta, inserting on a small protuberance and representing the outer seta of the baseoendopodite; the two inner setae small, the innermost one bare, both belonging to the exopodite. Inner lobe with two spinulose setae belonging to the endopodite, innermost seta two times as long as outer one.

Genital field as in Fig. 14 C.

Etymology. The species is named after the late Prof. Dr. W. Noodt in remembrance of his indispensable merits in the field of harpacticoid taxonomy.

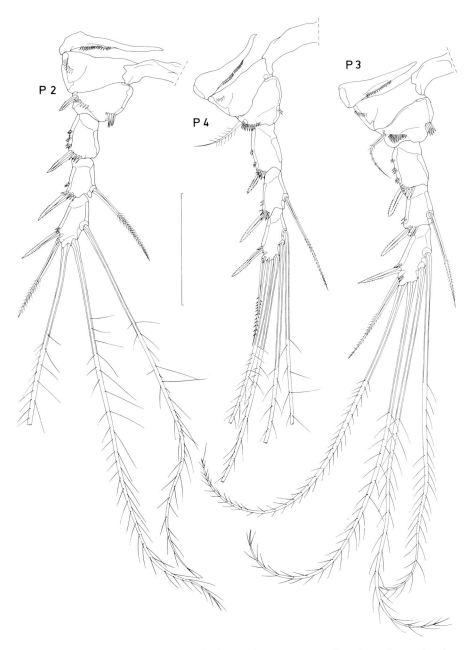

Fig. 17. *Talpina noodti* sp. n. Female. Right thoracic locomotor appendages P 2–4 in anterior view. Scale bar 50 μm.

Talpina fodens sp. n.

(Figs. 18–21)

Material. One female collected during Ant V/3 by AGT 10/590 (13. 11. 86) from about 400 m depth (K – 35098).
Locality 74° 54,7' S – 25° 58,5' W

Female
Body length 1400 (1365) μm and maximum body width 290 μm.
General body shape and morphology as in *T. noodti*.
Rostrum (Fig. 18) clearly bifid and plough-like, bearing 2 setae on ventral surface.
Caudal rami (Fig. 21) similar to that of *T. noodti*, but with one pinnate outer terminal spine instead of a bare seta. Surface of rami furnished with a few patches of spinules.
Antennula (Fig. 18) short and six-segmented, very similar to that of *T. noodti*. Segment II with one spinulose seta and five distal spines, one pinnate, and a spine-like pinnate seta on a process of posterior surface. Segment IV with a distal inner process carrying a long aesthetasc accompanied by a shorter bare seta. Aesthetasc of segment IV with a short subterminal spinule.
Armature formula: 1.–7.–5.–2.+Ae.–1.–10+Ae.
Antenna (Fig. 18). Coxa with sparse spinular ornamentation at distal inner corner. Allobasis bearing a plumose seta attached to proximal inner margin and a proximal one-segmented small exopodite. Exopodite with short pinnate terminal seta and a subterminal row of spinules. Endopodite almost as long as allobasis. Inner margin furnished with three strong spines and two sets of small spinules, the distal one consisting of three spines. Terminal edge with four non-geniculate setae, innermost one is shortest, next to outermost is longest. The two outermost setae are bipinnate and accompanied by a set of three short spinules at base.
Labrum (Fig. 19) of trapezoid shape and richly ornamented on posterior surface. Ornamentation arranged symmetrically and consisting of distal and lateral spinule rows, superficial spinular patches and combs, the latter associated with cuticular ridges. Medioproximal pore bearing bag-like tube; similar pore opening at the same position on anterior surface. Caudal edge anteriorly fringed with a semi-circular row of eight larger spinules.
Mandible (Fig. 19) with four strong spines on cutting edge, all bidentate. Proximal edge with two pinnate spines, the inner bipinnate one blunt and almost two times as long as adjacent one. Gnathobase ornamented with a few sets of spinules, one of which is situated on cutting edge, another one proximal to the pinnate spines. Palp with basis carrying a bipinnate seta on an inner process and

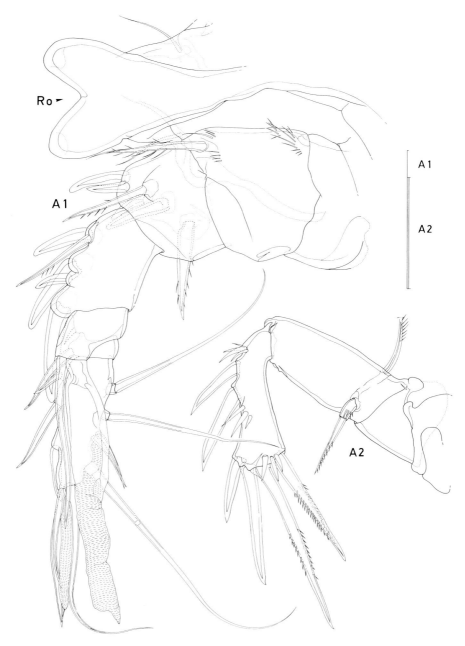

Fig. 18. *Talpina fodens* sp. n. Female. Antennula and rostrum in ventral view; left antenna in anterior view. Scale bar 50 μm.

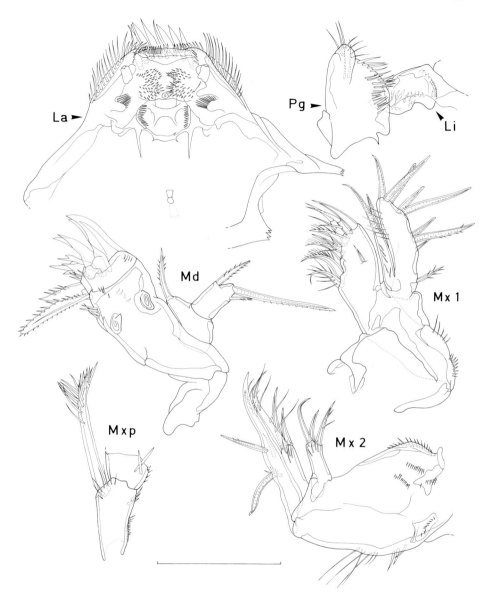

Fig. 19. *Talpina fodens* sp. n. Female. Labrum in posterior view; paragnath and lingua; mandible; maxillula; proximal portion of maxilliped; maxilla. Scale bar 50 μm.

with one-segmented endopodite bearing a relatively long pinnate seta terminally and two sub-terminal shorter setae of spine-like shape. Exopodite absent.

Paragnaths (Fig. 19) with distal lobe and richly ornamented on the inner surface. Anterior surface with peculiar comb of long spinules inserting on a cuticular ridge.

Lingua (Fig. 19) of scale-like shape and furnished with patches of spinules anteriorly.

Maxillula (Fig. 19). Precoxal arthrite broad with eight inwardly curved spines terminally, two of them are geniculate carrying one or two long spinules at base. Distalmost spine is bare, all others except the articulated ones are intensely spinulated. Posterior surface of arthrite bearing a single short seta. Endite of

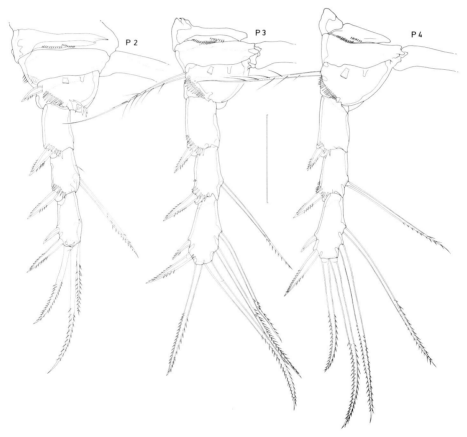

Fig. 20. *Talpina fodens* sp. n. Female. Thoracic locomotor appendages P 2–4 in anterior view. Scale bar 50 μm.

coxa slender, five times as long as wide, slightly tapering at tip and fringed with some spinules along posterior surface. Distal edge with two spine-like bare setae of different length. Basis with three terminal bare spine-like setae almost of the same length and a proximal clear pointed plumose spine. Distal part of inner margin with setule row. Endopodite is represented by two bare spine-like setae, the proximal one more than two times as long as the other. Exopodite reduced to a pinnate spine-like seta accompanied by two short bare ones.

Maxilla (Fig. 19). Syncoxa proximally with rich spinular ornamentation. Two slender endites of similar shape, each carrying a subterminal set of spinules and three terminal, inwardly curved setae, one of them spinulose. Basis terminally elongated into a claw-like process with a setular fringe along middle third. Two

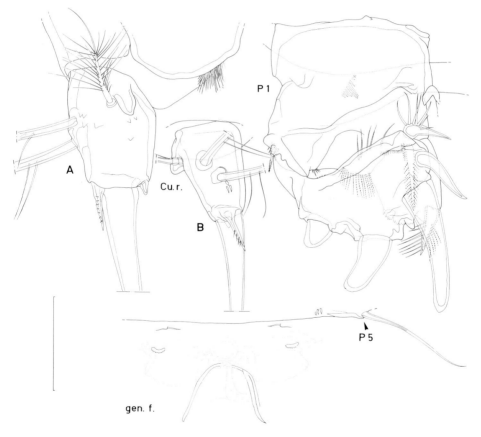

Fig. 21. *Talpina fodens* sp. n. Female. Left caudal ramus in dorsal view (A) and right caudal ramus in lateral view (B); left P 1 in posterior view and left side genital field with P 5. Scale bar 50 µm.

spine-like setae at base of claw. Endopodite represented by a bare seta. Exopodite represented by a bare strong seta with two smaller spinules at base.

Maxilliped (Fig. 19). Basis similar to that of *T. noodti*. Endopodite destroyed during preparation.

P 1 (Fig. 21) similar to that of *T. noodti* but blunt spines of exopodite less chitinized.

P 2–P 4 (Fig. 20) uniramous. Intercoxal plates slightly arched. Precoxae well developed and of triangular shape furnished with two distinct rows of tiny spinules along middle third of base line. Coxa of P 2 with oblique spinule row on outer edge of posterior surface. Basis with short pinnate spine on outer margin in P 2, while a long spinulose seta is present in P 3 and P 4. Proximal part of basis with pore bearing a bag-like tube. Oblique spinular row on anterior surface above insertion of exopodites. Endopodite in P 2 possibly represented by a fan of five spinules situated at distal inner protuberance. Exopodite with three segments of subequal length and size. Setae of terminal segment of P 2 shorter than those of P 3 and P 4, all bipinnate and shorter than in *T. noodti*. Spine and seta formulae as in *T. noodti*.

P 5 (Fig. 21) consisting of a narrow scale with a long bare seta representing outer seta of baseoendopodite.

Genital field as depicted in Fig. 21.

Etymology. The species is named after a presumed behavioural characteristic of *Talpina* and is derived from the Latin *fodere*, meaning to burrow.

Talpina furcispina sp. n.

(Figs. 22–26)

Material. One adult and one C IV female collected during Ant VII/4 by AGT 14/295 (21. 2. 89) from 2000 m depth (K – 35099).
Locality. 71° 08,8' S – 13° 48,1' W

Female

Body length 1630 (1550) μm and maximum body width 200 μm. Body (Fig. 22) vermiform. General body shape and morphology similar to that of *T. noodti*. Last urosomal somite ventrally with spinule comb and intense setule trimmings next to caudal rami.

Rostrum (Figs. 22, 23) prominent and plough-like, carrying a pair of setae on anterior surface; flanks wing-like. Posterior surface medially with a pair of small tube pores.

Caudal rami (Fig. 26) 2.5 times as long as wide and bent medially, carrying five plumose setae. Dorsal one biarticulated at base, inserting next to a small

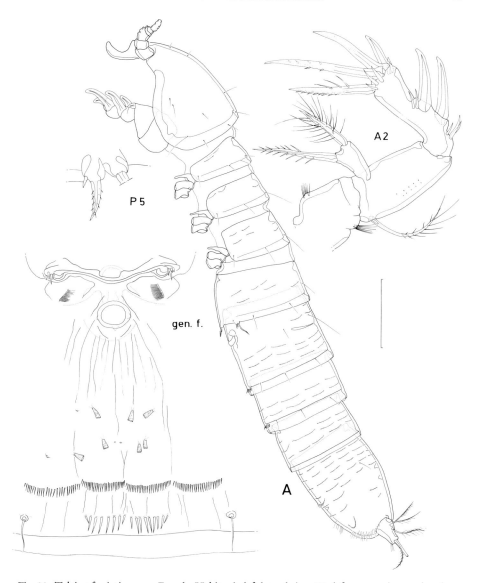

Fig. 22. *Talpina furcispina* sp. n. Female. Habitus in left lateral view (A); left antenna in anterior view; left P 5 baseoendopodite and exopodite and genital field. Scale bar 50 μm and 200 μm for appendages and genital field/habitus respectively.

inner lobe. Two outer setae slightly shifted dorsally. Distal inner corner bearing a strong spiniform emargination with pore at tip and is furnished with five long spinules ventrally. Dorsal surface of rami with two pores, one adjacent to biarticulated seta, the other distal to lateral setae bearing a long tube.

Antennula (Fig. 23) short and six-segmented. Segment I with two setule fans at inner edge and spine-like plumose seta at distal inner corner. Segment II with six spiniform elements surrounding caudal segment border in a semi-circular manner. Outermost spine is the largest one and bipinnate like the innermost one. Segment III carrying three strong spines on inner margin accompanied by a seta and a minute crooked spine. Segment IV small with aesthetasc and seta on the distal inner process. Aesthetasc with terminal spinule and chitin supported surface. Seta less than half as long as aesthetasc. Segment V small with a single spine at distal inner corner. Segment VI bearing five long articulated setae on the outer margin, two spines on inner margin, and a terminal complex of two setae and a short aesthetasc confluent at base. This aesthetasc heavily chitinized at base. Armature formula: 1.–6.–4.–1+Ae.–1.–9.+Ae

Antenna (Fig. 22). Coxa bearing two setule fans on both lateral margins. Allobasis proximally with one inner plumose seta. Exopodite one-segmented, distally widened and with two terminal setae of similar length; one of which bipinnate, the other plumose. Second segment of endopodite with one small proximal and two strong spines midlength on inner margin and two sets of small spinules along inner margin. Terminal edge carrying four nongeniculate setae increasing in size from innermost to outermost one. The two outermost setae bipinnate and with three small spinules at base.

Mandible (Fig. 24). Cutting edge with a blunt distal process and an adjacent bidentate spine; other teeth small. Proximal edge with two heavily pinnated setae, the proximalmost backwardly curved and two times as long as the other one which is inwardly curved. A set of spinules is situated distally as well as proximally to those two setae. Palp consisting of well developed basis carrying a plumose seta at distal inner corner and a slightly curved endopodite with three bipinnate terminal setae, the middle one longest. The ornamentation of this seta is composed of long spinules along the middle third and of substantially smaller ones along the distal third. Exopodite absent.

Maxilla (Fig. 24). Syncoxa proximally with two spinule fans on both sides. Two endites of cylindrical shape with three terminal setae and a subterminal setule fan each. Proximal endite with two spinulose setae. Basis terminating into a claw serrate along distal half. Two setae attached at base of claw. Endopodite appearing as a single bare seta. Exopodite absent.

Maxilliped (Fig. 26) prehensile. Basis ornamented with several sets of spinules including a large fan at the base of the distal inner spiniform seta which is

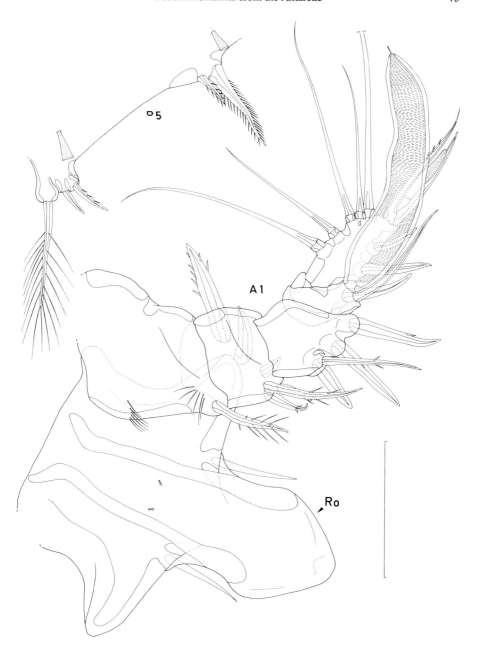

Fig. 23. *Talpina furcispina* sp. n. Female. Right P 5; antennula and rostrum in ventral view. Scale bar 50 μm.

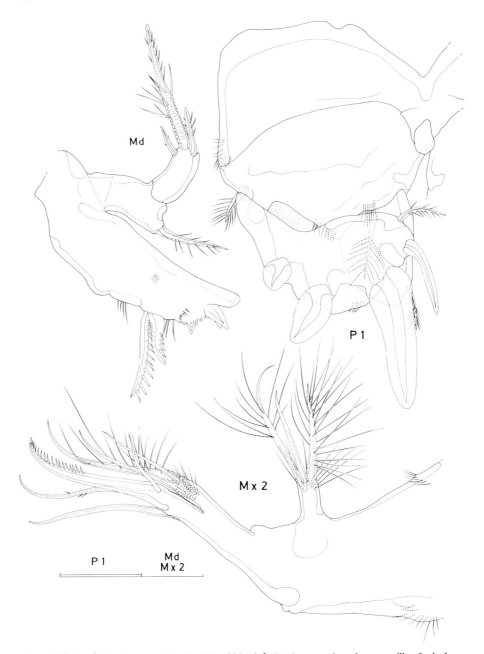

Fig. 24 *Talpina furcispina* sp. n. Female. Mandible; left P 1 in posterior view; maxilla. Scale bars 50 µm.

furnished with long spinules at tip. First segment of endopodite fringed with long spinules along inner margin. Outer margin and anterior surface with fans of long spinules as well. Second segment of endopodite small carrying a basal minute spine and a terminal claw-like seta serrate along the two distal thirds.

P 1 (Fig. 24) similar to that of the above described new species of *Talpina*. Endopodite with smooth outer edge and slender subterminal plumose seta.

P 2–P 4 (Fig. 25) biramous. Intercoxal plates and precoxae similar to those of the above described species. Coxae without any ornamentation. Bases with a

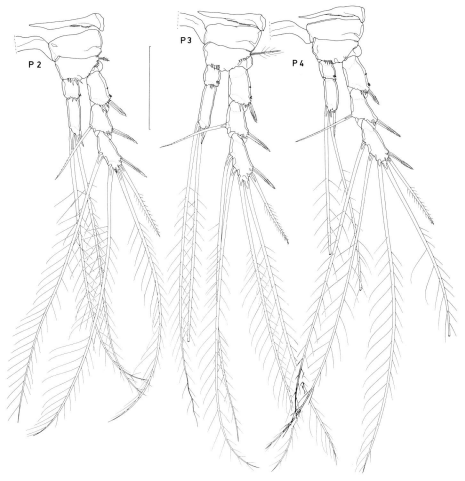

Fig. 25. *Talpina furcispina* sp. n. Female. Left thoracic locomotor appendages P 2–4 in anterior view. Scale bar 50 µm.

spinule row above the insertion of endopodite and oblique spinule row near outer margin, the latter absent in P 4. Outer seta of basis in P 2 bipinnate and spiniform, in P 3 long and plumose; lacking in P 4. Endopodite two-segmented. Proximal segment almost square, terminal segment cylindrical and about two times as long as proximal one. Exopodite with three segments of subequal length and size. Terminal and inner setae of both rami long and intensely plumose.

Spine and seta formulae as follows:

	Exopodite	Endopodite
P 2	0.1.212	0.021
P 3	0.1.222	0.021
P 4	0.1.222	0.021

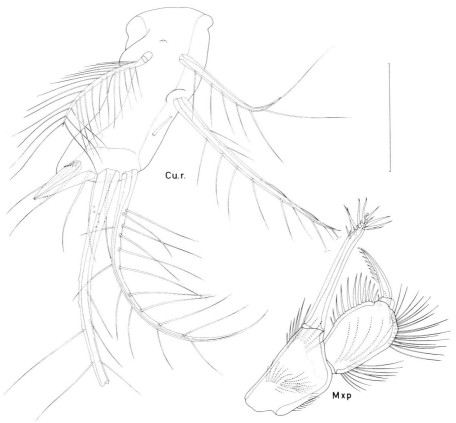

Fig. 26. *Talpina furcispina* sp. n. Female. Right caudal ramus in dorsal view; maxilliped. Scale bar 50 μm.

P 5 (Fig. 23) similar to that of *T. noodti* but revealing a striking asymmetry in the structure of the exopodite concerning the considerably reduced innermost setae on the left exopodite (Fig. 22). A pore with a long bag-like tube is associated with each lobe.

Genital field as depicted in Fig. 22.

Etymology. The species name refers to the morphological characteristic of bearing an inner spine of the caudal rami, derived from the Latin *furca* and *spina*, meaning fork and thorn respectively.

Talpina bathyalis sp. n.
(Figs. 27–30)

Material. One female collected during Ant VIII/6 by GKG 16/536, sieving the supernatant water of the box corer (2. 4. 1990) from 3970 m depth (K – 35100).

Locality. 66° 55,08' S – 34° 18,20' E

Female

Body length 790 (750) µm and maximum body width 140 µm.

Body (Fig. 27) vermiform. Surface of all somites except cephalosome ornamented with a peculiar pattern of spicule rows and patches. Long spinules laterally and dorsolaterally on the posterior margins of thoracal somites, genital somite, and third urosomal somite. Ultimate urosomal somite ventrally with a paired subterminal row of spinules. General morphology of the body resembling that of the species described above.

Rostrum (Fig. 28) prominent but not as plough-like as in the above described species. Tip recurved. Paired setae on anterior surface.

Caudal rami (Fig. 30) similar to those of *T. noodti* and *T. fodens* but even the terminal outer seta reduced to a minute spine. Dorsal surface with pore bearing a bag-like tube.

Antennula (Fig. 28) short and six-segmented. Segment I with a bipinnate seta at the distal inner corner. Segment II bearing six armature elements arranged in an oblique semi-circle along the distal segment edge. Seta on outer edge stout and plumose. Segment III with three spines and two proximal setae along inner margin. Segment IV short with process at distal inner corner carrying an aesthetasc and one seta confluent at base and one small seta proximally. Segment V small with a single outer spine-like seta. Segment VI terminally with a short aesthetasc bearing a minute spinule at tip. Aesthetasc accompanied by three bare setae. Outer margin of segment with five articulated setae, the proximalmost and distalmost one long, the others short and spiniform. Inner margin furnished with three spines.

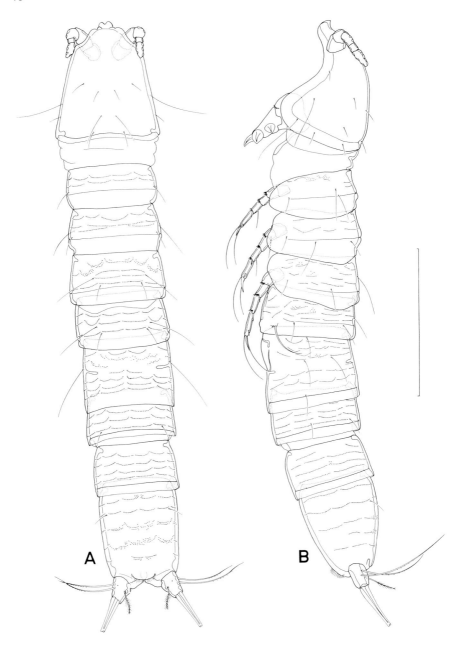

Fig. 27. *Talpina bathyalis* sp. n. Female. Habitus in dorsal (A) and left lateral view (B). Scale bar 200 μm.

Armature formula: 1.–6.–5.–2 + Ae.–1.–11 + Ae.

Antenna (Fig. 28) with unornamented coxa. Allobasis proximally with one plumose seta on the inner margin. Exopodite absent. Endopodite with six stout spines, two of which on inner margin; in other species there is a third proximal spine along the inner margin – this is absent here but its former location is indicated by a row of spinules. The other ones terminally and nongeniculate. Outermost terminal spine bipinnate, next to outermost one is longest and bipinnate as well. Three sets of spinules, two of which accompanying the proximal spines, the third one is situated at base of terminal setae.

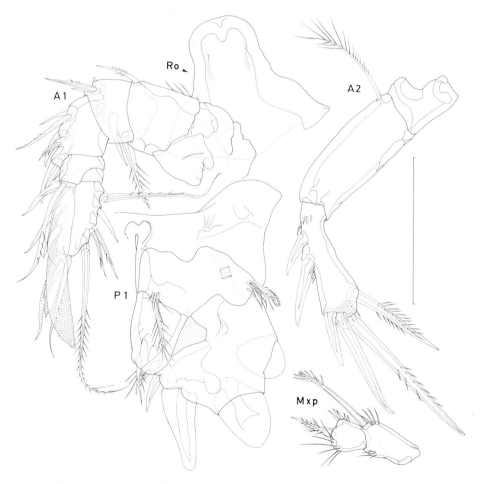

Fig. 28. *Talpina bathyalis* sp. n. Female. Antennula and rostrum in ventral view; right antenna in posterior view; left P 1 in anterior view; maxilliped. Scale bar 50 μm.

Maxilliped (Fig. 28). Basis with distal sets of spinules. Long spine at distal inner corner bearing several long spinules along terminal third. First segment of endopodite with a set of stout spinules subdistally on the inner protuberance and a fan of long spinules at distal outer corner. Second segment of the endopodite

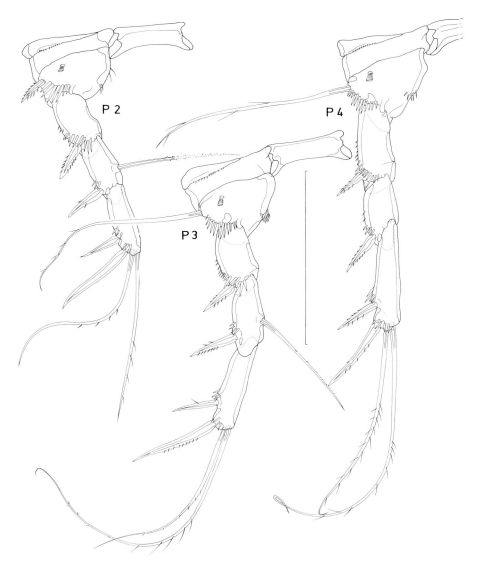

Fig. 29. *Talpina bathyalis* sp. n. Female. Right thoracic locomotor appendages P 2–4 in anterior view. Scale bar 50 μm.

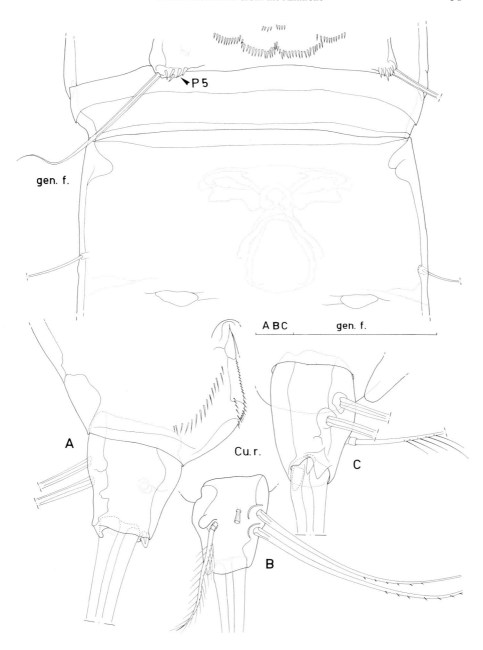

Fig. 30. *Talpina bathyalis* sp. n. Female. Genital field; P 5; caudal ramus in ventral (A), dorsolateral (B) and lateral view (C). Scale bars 50 μm.

fused with the terminal bipinnate seta which is indicated by the thickening at base.

P 1 (Fig. 28) similar to that of *T. furcispina* but smaller. Posterior surface of basis with pore bearing a bag-like tube. Exopodite wider at base than in the species described above.

P 2–P 4 (Fig. 29) uniramous. Intercoxal plates of trapezoid shape with lobe-like extended angles. Base line slightly bent inwards. Precoxae triangular with up to 13 tiny spinules arranged in a row above insertion of the coxa. Coxae unornamented, rectangular, and four times as wide as long. Bases with characteristic tube pore on anterior surface and long outer seta which is developed as a short, stout, and bipinnate spine in P 2. Oblique row of stout spinules above insertion of the exopodites. Endopodites possibly represented by a few spinules on the inner margin of bases. Exopodites three-segmented, first and last segment increasing in length from P 2 to P 4. Terminal setae slender and only sparsely spinulated. That of P 2 shorter than in the other thoracopods. Spine and seta formulae as follows:

	Exopodite	Endopodite
P 2	0.1.022	–
P 3	0.1.022	–
P 4	0.0.022	–

P 5 (Fig. 30) largely reduced, consisting of a small lobe with a long, bare outer seta and a set of five stout spinules.

Genital field as depicted in Fig. 30.

Etymology. The species name refers to the deep sea occurrence of the species, derived from the Greek *bathys*, meaning deep.

Talpina pectinata sp. n.

(Figs. 31–35)

Material. One male collected during Ant VII/4 by MG 14/277 (16. 2. 89) from about 400 m depth (K – 35101).

Locality. 71° 40,0' S – 12° 35,9' W

Male

Body length 1150 (1100) µm and maximum body width 210 µm.

Body (Fig. 31 A) vermiform. Cephalosome and thoracal somites dorsally and laterally furnished with long spinules. Denticle row patterns especially marked at urosome. Urosomal somites ventrolaterally with a double spinule fan on posterior margin (Fig. 33 A). Ultimate abdominal somite additionally with two rows of strong spinules including multidentate ones on ventral surface forming a

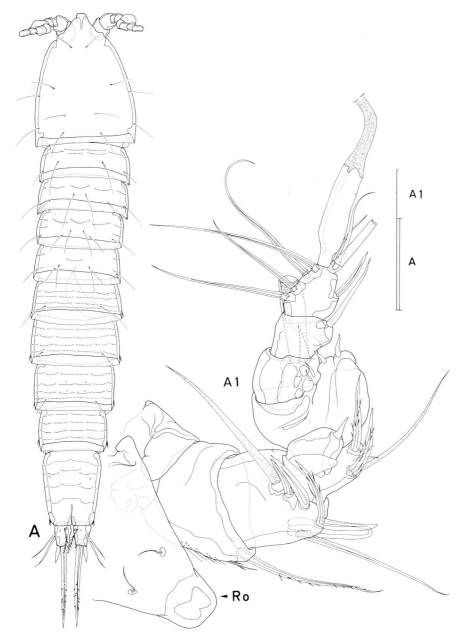

Fig. 31. *Talpina pectinata* sp. n. Male. Habitus in dorsal view (A); antennula and rostrum in dorsal view. Scale bars 50 μm.

comb (Fig. 35 A). Paired peculiar rods (tube pores?) subdistally on lateral surface of anal somite.

Rostrum (Fig. 31) triangular, plough-like and duck bill-shaped at tip with a pair of setae on dorsal surface.

Caudal rami (Fig. 35) subcylindrical, almost two times as long as wide. Dorsal seta plumose and biarticulated at base inserting close to a small lobe on inner margin. Two dorsolateral setae of equal length furnished sporadically with spinules. Terminal seta short (180 μm in length) and stout, with denticles along third quarter. Strong spine on outer terminal edge. Inner edge with minute spine. Ventral surface terminally furnished with one or two strong spinules. Pore with bag-like tube on dorsal surface.

Antennula (Fig. 31) seven-segmented, subchirocer. Segment I with several spinules along inner margin and pinnate seta at distal inner corner. Segment II distally with seven elements, of which two are long bare setae and one is a short spinulose one. Segment III bilobed. Outer lobe with a single terminal spine, inner one carrying two pinnate spines, a bare long seta and a smaller spine. Segment IV swollen with two terminal spines at inner distal corner. Dorsal process with spine and large aesthetasc supported by a cuticular ledge. Segment V with two small spinules. Segment VI bearing a single stout bare seta at distal inner corner. Segment VII sub-terminally with a long aesthetasc chitinized at base and the proximal half, accompanied by two bare setae. At its base there is a large articulated seta. Outer margin furnished with four long setae articulated at base, inner margin with one single spine.

Armature formula: 1.–7.–5.–3 + Ae.–2.–1.–8. + Ae.

Antenna (Fig. 32). Coxa unornamented. Allobasis, proximally with one long plumose seta on inner margin; former distal borderline of basis indicated. Exopodite small, one-segmented with pinnate spiniform seta. Endopodite 2-segmented, first unarmed and second terminally with four non-geniculate more or less bipinnate setae, the innermost one is shortest, the next to outermost one is longest. Two spines proximally on the inner margin. Three sets of spinules, two on inner margin, the third terminally.

Mandible (Fig. 32) without genuine cutting edge. *Corpus mandibulae* elongate and tapering to a point. Two setae on inner surface. Palp two-segmented. Basis well developed with one seta at distal inner corner. Endopodite cylindrical with terminal long seta accompanied by a shorter one and a pinnate spine. Exopodite absent.

Maxillula (Fig. 32). Coxa with two small spinules replacing the endite. Basis slender, almost four times as long as wide, slightly curved, four bare setae distally. Endopodite represented by two bare setae distal to a smaller one. Exopodite appearing as a single spinulose seta.

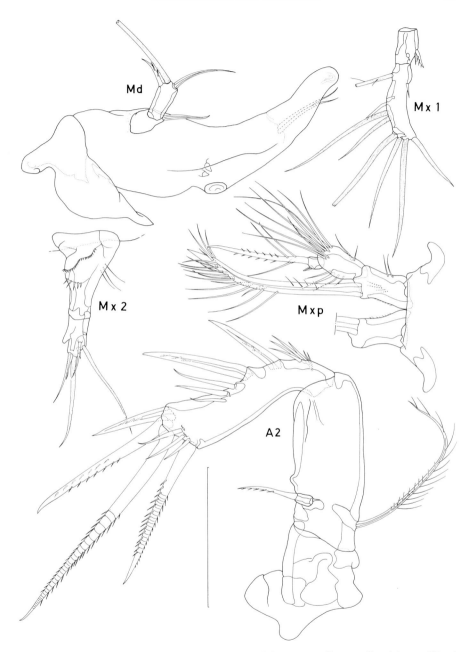

Fig. 32. *Talpina pectinata* sp. n. Male. Mandible; maxillular protopodite; maxilla; right maxilliped; left antenna in posterior view. Scale bar 50 μm.

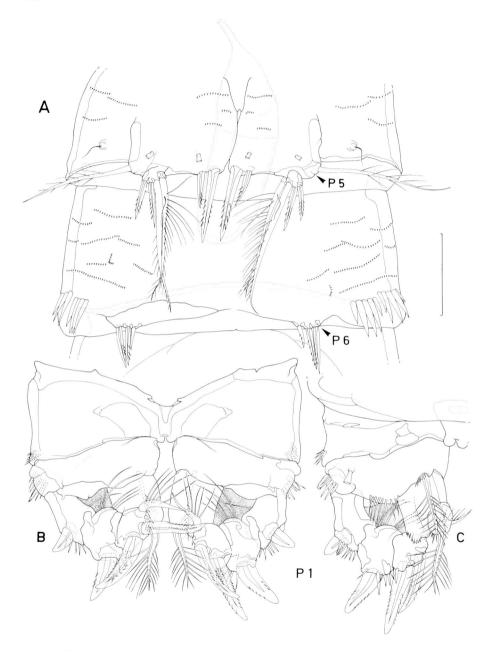

Fig. 33. *Talpina pectinata* sp. n. Male. Hind appendages P 5 and P 6 in ventral view (A); pair of P 1 in posterior view (B); left P 1 in anterior view. Scale bar 50 μm.

Maxilla (Fig. 32) reduced to a cone-like appendage without endites. Syncoxa with spinular ornamentations. Basis with a small spine and five setae representing the terminal claw and endopodite respectively.

Maxipilled (Fig. 32) well developed. Basis with long inner spine bearing several long spinules along distal half. First segment of endopodite with long spinules at distal outer corner. Inner margin sparsely fringed with spinules. Second segment of endopodite carrying a subdistal small spine and a terminal long claw-like seta, bipinnate along middle third and proximally furnished with long spinules. Spine of basis longer than basis and endopodite including terminal claw together.

P 1 (Figs. 33 B, C). Coxae of both counterparts completely fused carrying a spinule fan anteriorly on distal outer margin. Basis posteriorly with plumose seta and spinular fan above insertion of endopodite, oblique spinule row above insertion of exopodite, and setule row between the insertions of both rami. Pore with bag-like tube on anterior surface. Exopodite three-segmented and inwardly flexed. Terminal segment is the smallest one. All segments with one spiniform seta each on outer margin increasing in size from first to last segment. Last segment additionally with terminal spine, small remnant of spine and inner plumose seta. Spines of last two segments bipinnate. Outer margin of first two segments distally fringed with spinules. Endopodite two-segmented. First segment as an almost square plate with toothed outer distal corner. Second segment small bearing a plumose seta terminally.

P 2 – P 4 (Figs. 34, 35) biramous. Intercoxal plates arched with lateral protuberances. Precoxae scale-like with a distal spinular row. Coxae unornamented. Bases with outer seta, spinular row on outer margin, and spinule fan between insertions of exopodite and endopodite; pore with bag-like tube on anterior surface. Exopodite three-segmented. Segments decreasing in size and length from proximal to distal. Last segment of P 3 bearing a pore with baglike tube on anterior surface. Endopodite two-segmented in P 2 and P 4. First segment squat, second one long and slender, but reduced in P 4. Subdistal outer spinulose seta of last segment in P 2 – 4 modified. In P 4 it is a distally swollen heavily bipinnated element. Endopodite of P 3 three-segmented. First segment as in P 2 and P 4. Second segment small with a strong and modified seta exceeding the distal edge of third segment. Spine and seta formulae as follows:

	Exopodite	Endopodite
P 1	0.0.121	0.0.010
P 2	0.1.122	0.221
P 3	0.1.222	0. mod. 020
P 4	0.1.222	0.221

P 5 (Fig. 33 A) plate-like. Baseoendopodite with outer plumose seta and two bipinnate spiniform setae at inner distal corner. Two pores with bag-like tube. Exopodite consisting of small lobe partially fused with baseoendopodite, bearing two long inner plumose setae and two outer bipinnate spiniform setae.

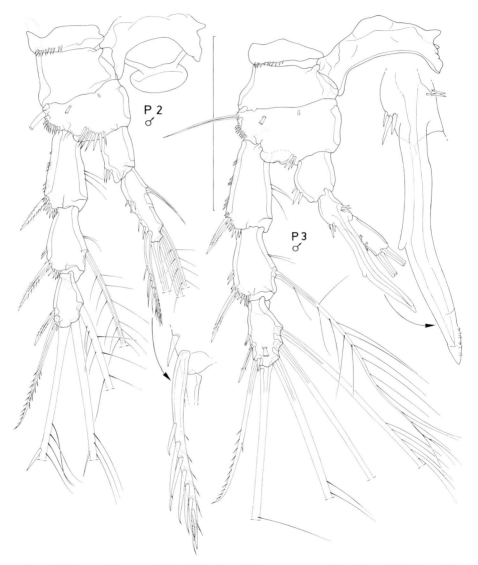

Fig. 34. *Talpina pectinata* sp. n. Male. Right thoracic locomotor appendages P 2 and P 3 in anterior view. Scale bar 50 μm.

Metahuntemannia from the Antarctic 59

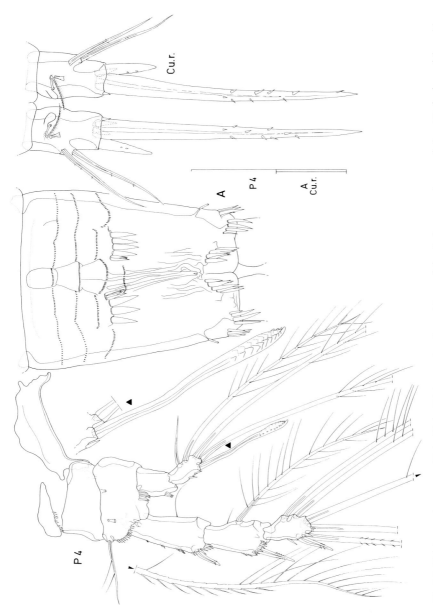

Fig. 35. *Talpina pectinata* sp. n. Male. Right P 4 in anterior view; ultimate abdominal segment in ventral view (A); both caudal rami in dorsal view. Scale bar 50 μm.

P 6 (Fig. 33 A) as a narrow lobe, distinctly separated only on the right side (asymmetrical), bearing three pinnate setae.

Spermatophore (Figs. 31 A, 33 A) elongate and with a short neck.

Etymology. The species name refers to the morphological characteristic of the ventral abdominal somites bearing rows of large spinules ventrally and ventrolaterally; derived from the Greek *pecten*, meaning comb.

Key for the identification of species belonging to *Metahuntemannia* and *Talpina*:

1. – Body fusiform (if vermiform than caudal setae reduced in size);   2
   – Body vermiform; Ro plough-like; P 1 exp transformed and bent inwardly in the females   18
2. – P 2–4 enp 2-seg   3
   – P 3 enp 3-seg   8
3. – P 1 exp terminal seta bent inwardly   4
   – P 1 exp terminal seta straight   5
4. – A 2 allobasis with 1 proximal and 1 distal seta   *M. crassa*
   – A 2 allobasis with 1 proximal seta   *M. mediterranea*
5. – P 5 exp and enp fused to a plate   6
   – P 5 exp and enp subdivided by an invagination   7
6. – A 2 exp absent; P 5 plate in total with 7 setae   *M. iberica*
   – A 2 exp 1-seg with 1 seta; P 5 plate in total with 6 setae   *M. magniceps*
7. – Prospective P 5 enp with 3 setae   *M. spinipes*
   – Prospective P 5 enp with 2 setae (both sexes); Mxp mostly transformed and being unornamented and lacking seta of syncoxa   *M. pseudomagniceps*
8. – P 2, 4 enp 2-seg; P 3 enp 3-seg   9
   – P 2–4 enp absent   11
   (except P 2 enp 2-seg in *M. beckeri*)
9. – A 2 exp with 1 seta   10
   – A 2 exp with 2 setae   *M. dovpori*
10. – P 1 exp III male with 2 inner setae   *M. triarticulata*
    – P 4 enp with 1 inner seta   *M. smirnovi*
    (as in the female of *M. triarticulata*)
11. – A 2 exp absent   12
    – A 2 exp present (1-seg with 1 seta)   13
12. – Body fusiform; Mxp syncoxa with 2 setae   *M. spinosa*

	– Body vermiform; caudal setae reduced in size	*M. indica* (syn. *B. indica*)
13.	– P 1 exp III with 2 spines; Md enp separate	14
	– P 1 exp III with 2 spines and 1 inner seta; Md enp fused with basis; A 2 allobasis without distal seta	*M. arctica*
14.	– P 2–4 exp III with 3 setae	15
	– P 2–4 exp III with 4 setae	16
15.	– A 1 III with 1 seta and 3 spiniform elements	*M. drzycimskii*
	– A 1 III with 2 setae and 3 spiniform elements	*M. texturata*
16.	– A 2 allobasis with 2 setae on the inner margin	17
	– A 2 allobasis with 1 seta on the inner margin (this character is not explicitly stated in the text by SMIRNOV 1946)	*M. gorbunovi*
17.	– A 1 II with 1 seta	*M. atlantica*
	– A 1 II with 3 setae	*M. beckeri*
18.	– P 2–4 enp reduced;	19
	– P 2–4 enp at least 2-seg;	23
19.	– A 2 exp with 1 seta; P 1 enp 1-seg with 1 plumose seta	20
	– A 2 exp with at least 2 setae; P 1 enp 1-seg with 1 thorn-like element	21
20.	– P 2–4 enp remnants as spinule crests on basis; 1 bag-like tube-pore on cu.r.	*T. noodti*
	– P 2–4 enp completely absent	*T. fodens*
21.	– Md enp with 1 seta; Md cutting edge complex	*T. bifida*
	– Md enp with 2 setae; Md cutting edge two-tipped, dagger-shaped	22
22.	– A 2 exp with 3 setae	*T. curticauda*
	– A 2 exp with 2 setae	*T. talpa*
23.	– A 2 exp with 2 setae	24
	– A 2 exp with 1 seta	25
24.	– P 2–4 enp 2-seg	*T. furcispina*
	– P 2, 3 enp 3-seg; P 4 enp 2-seg	*T. micracantha*
25.	– Md cutting edge dagger-shaped; ultimate abdominal segment with large spinules ventrally	*T. pectinata*
	– Md cutting edge complex	27
27.	– P 1 basis and enp fused; P 5 exp separate in the males	*T. peruana*
	– P 1 basis and enp separate; P 5 exp fused with baseoendopodite	*T. pacifica*

D. Discussion

As BECKER (1979) has pointed out already, *Metahuntemannia* is remarkably adapted to an endopelic existence. According to this author especially the rostrum, labrum, P 1 and habitus are subjected to a remarkable transformation in a synorganised fashion within the group under consideration.

The rostrum is a small triangular plate in *Metahuntemannia* (and *Beckeria* Por, 1986 a synonymized herein as *Metahuntemannia*, see below) whereas it becomes plough-like in *Talpina*. A plough-shaped labrum is developed in *M. iberica* and *M. magniceps* demonstrating the functional demand of such a structure in the frontal portion of the body, which is developed from very different structures. The P 1 exp is remarkably transformed, especially in the females. From a straight 3-segmented ramus in *Metahuntemannia* it is bent inwardly in *Talpina*, becomes 1-segmented due to the fusion of its segments in the bulk of its representatives, and even a triangular plate bent to almost a right angle in *T. peruana*. The P 1 enp is represented by a thorn-like element in *Metahuntemannia* and a flap-like segment in *Talpina*. In *H:a temannia* the endopodite is 2-segmented bearing 2 setae, it is 1-segmented in *Talpina* and represented only by a thorn-like element in *Metahuntemannia*. Whether this thorn of the latter taxon represents the transformed endopodal segment, or rather the armature element of the single segment, is difficult to decide. In one representative of the related *Talpina*, i. e. *T. peruana*, however, a fusion has taken place in the females even of the basis and endopodite of the P 1 (cf. character 53, Fig. 39). This fusion is a peculiarity, and, to our knowledge not known from any other harpacticoid leg than the P 5.

From a fusiform habitus in *Metahuntemannia* a vermiform body form has evolved in *Talpina* as an adaptation to a burrowing life style. The above mentioned characters are the autapomorphies of *Talpina* representing a peculiar "Lebensformtypus" especially among the females of this taxon.

In his thesis BECKER discussed the tendency of reduction in the maxilliped as is demonstrated by *T. curticauda*. However, this cannot be correlated with any evolutionary trend of other appendages. His opinion that a reduced maxilliped is characteristic for the *spinosa*-group (i. e. *Metahuntemannia* s. str.) cannot be supported here taking all the species into consideration described in the meanwhile.

The trend for the reduction (and/or secondary acquisition) of the P 2–4 endopodites is pronounced among both, *Metahuntemannia* and *Talpina*. The character analysis, however, has shown that the reduction of the P 2–4 endopodites must have taken place for more than once within unrelated lineages independently. Therefore, no interspecific morphocline is demonstrated. However, as suggested by BECKER (1979), it can be stated that there is a tendency for the reduction of segments from rear to front. In this context it is interesting to note that there are either none, 2 or 3 segments on the P 2–4 endopodites of *Metahuntemannia* and *Talpina* but never 1 segment as in *Huntemannia* spp.

As was pointed out by BECKER (1979) already, there is no other taxon within Harpacticoida with a similarly pronounced degree of sexual dimorphism than *Metahuntemannia*. Although there is a lack of information especially on male morphology (see discussion below) from the material available the following can be stated.

As in other representatives of Harpacticoida the females are much larger than the males (Table 1). Not only the antennules, P 2–4 endopodites, P 5 and P 6 which used to be sexually dimorphic in other harpacticoids are sexually modified here. In both taxa, *Metahuntemannia* and *Talpina*, also the P 1 morphology is different and less transformed in the males. Additionally, the maxilliped and even the maxilla is much reduced in the females of both taxa. A sexually dimorphic maxilla has otherwise not been reported for the Harpacticoida yet. Dimorphic maxillipeds, however, are well known from Tisbidae (cf. DAHMS et al. 1991). From no other harpacticoid a sexually dimorphic rostrum is known so far as the plough-like one in the females of *Talpina* which is triangular and not prominent in the males.

As was suggested by BECKER (1979) the females and males of one particular species may belong to quite different "Lebensformtypen" *sensu* REMANE probably performing different life habits. In addition to the sexually dimorphic characters mentioned above this becomes evident by the fusiform habitus of the females but vermiform habitus of the males of e. g. *T. talpa* and *T. peruana*. An explanation for this phenomenon may be the use of different microhabitats of the environment, e. g. surface-near or deeper horizons of the flocculent layer which becomes more sticky in deeper layers and may require better burrowing capabilities there.

Evolutionary states of characters used for the cladograms (Figs. 36–38)

The plesiomorphic states of characters 2–4 (Fig. 36) are deduced from the most related outgroup of *Metahuntemannia* and *Talpina*, i. e. *Huntemannia* Poppe, 1884 and from other presumably related taxa belonging to the Cletodidae (e. g. *Nannopus, Pontopolites, Pseudocletodes*).

A pair of egg-sacs carried by the females of *Huntemannia* (1) is otherwise only known from Canuellidae, Parastenocarididae (i. e. two single eggs) and Diosaccidae among Harpacticoida.

The autapomorphies (2–4) are remarkable peculiarities for *Metahuntemannia*

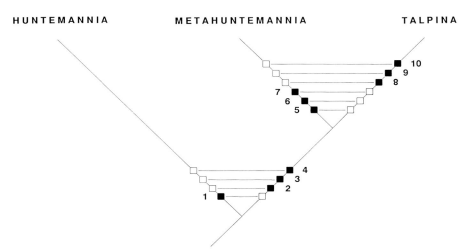

Fig. 36. Cladogram based on adult characters depicting the relationships of *Huntemannia*, *Metahuntemannia* and *Talpina* (a = apomorphic; p = plesiomorphic). Character states as follows:
1. Females bearing two egg-sacs (a); females with a single egg-sac (p)
2. Mxp reduced in size and seta of syncoxa transformed (a); Mxp ordinary and seta of syncoxa small (p)
3. P 1 enp transformed (a); P 1 enp ordinary and 1-seg (p)
4. Cu. r. with inner dorsal lobe (a); Cu. r. without lobe (p)
5. Cu. r inner dorsal lobe prominent and spinulated (a); Cu. r. lobe small and unornamented (p)
6. A 1 seg III with denticulated spine-like setae (a); A 1 setae not denticulated (p)
7. P 1 enp transformed to a thorn (a); P 1 enp 1-seg, lobe-like (p)
8. Habitus vermiform (a); habitus fusiform (p)
9. Rostrum plough-like; Md basis with 1 seta (a); rostrum triangular, Md basis without seta (p)
10. P 1 exp 1- or 3-segmented and bent inwardly in the females (a); P 1 exp 3-seg and straight (p)

and *Talpina* respectively. The maxillipeds of representatives belonging to both taxa are much less prehensile (2) compared to other stenopodial maxillipeds among Harpacticoida. The P 1 endopodites of other cletodid genera are not transformed to a degree (may it be lobe-like in *Talpina* or represented by a thorn in *Metahuntemannia*) as in the taxa under consideration (3). There is no dorsal lobe of the caudal rami known from other taxa of Cletodidae (4). Both, characters 5 and 6 are autapomorphies for *Metahuntemannia* for they are not known from any other cletodid. This holds especially for the 3 denticulated spiniform setae on A 1 III (6) which are present in all representatives of *Metahuntemannia* but in no other taxon of Cletodidae. *Talpina* is most characterized by the following autapomorphies not known from other taxa of Cletodidae: its ploughlike rostrum and mandibular basis bearing 1 seta (9) and the highly transformed sexually dimorphic P 1 with inwardly bent exopodites in the females (10), sometimes plate-like due to fusion of segments, and a lobe-like endopodite bearing a plumose seta. A vermiform habitus (8), however, has probably evolved within several independent lineages among Harpactioida adapted to an interstitial or endopelic life style (e. g. in Cylindropsyllidae, Neobradyidae and Parastenocarididae). However, in Cletodidae it serves as an autapomorphy for *Talpina*.

The taxonomy of the 17 species of *Metahuntemannia* is rather difficult and the cladogram presented, therefore, must be preliminary in large portions (Fig. 37). This is due on the one hand to the lack of character information from the literature (missing appendages, broken armature, low standard of drawings and descriptions) and on the other hand to the remarkable degree of sexual dimorphism of the taxa under consideration (see above). Most of the species are described on the base of one specimen representing one sex only. Thus, characters of both sexes have to be compared without knowing exactly the presence and degree of sexual dimorphism. However, in the following, character states are discussed in the light of morphological information which is available yet.

A 2-segmented endopodite of P 2–4 is not a strong autapomorphy for a group of 6 species belonging to *Metahuntemannia* for the fact that it may have developed for more than once (20). In *Talpina* even a total loss of the P 2–4 endopodites is proposed to have taken place for two times independently (see Fig. 38). Also, the proposed 3-segmented state as being plesiomorphic is questionable, for *Huntemannia* bears minute 1-segmented endopodites, but this may be a secondary acquisition for 3-segmented endopodites occur within *Metahuntemannia* as well as in *Talpina*.

The bending of the terminal seta of the P 1 exp (21) on the other hand is a convincing synapomorphy for *M. mediterranea* and *M. crassa* for it is restricted to these two species. In general, *M. crassa* (syn. *Apodella crassa* Por, 1965) and *M. mediterranea* are of remarkable similarity. This holds for all external characters including all the oral appendages. According to the descriptions, there are only differences in the setal armature of the 6-segmented antennules, the additional distal seta of the antennal allobasis in *M. crassa*, the shape of P 2–4 exopodite segments, possibly the shape of the caudal ramus, longer caudal setae other than the major terminal one in *M. crassa* and the appearance of the genital field. In case there really is a third (= terminal) spine of P 1 exp III in *M. crassa* as indicated by POR (1965) in hatched line (and the shape of the segment indicates this), this would establish the most striking difference to *M. mediterranea* counting only 2 spiniform setae on P 1 exp III.

The peculiar plough-like labrum (22) is a convincing synapomorphy for *M. iberica* and *M. magniceps* corresponding to the plough-like rostrum (9) which is constitutive for *Talpina*.

The lack of the antennal exopodite (23) is not a strong apomorphy for *M. iberica* for it occurs also in *M. spinosa*, *M. indica* and *T. bathyalis*. A P 5 with

METAHUNTEMANNIA

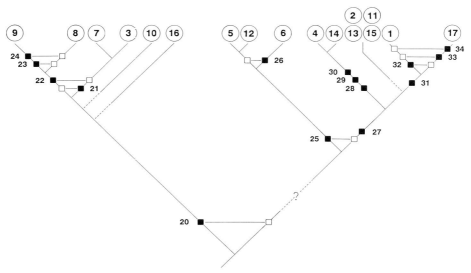

Fig. 37. Cladogram based on adult characters depicting the relationship within *Metahuntemannia* s. str. (a = apomorphic; p = plesiomorphic). Character states as follows (encircled numbers refer to the species as listed in Table 1; sole numbers refer to characters):
20. P 2–4 enp 2-seg (a); P 2–4 enp 3-seg (p)
21. P 1 exp terminal seta bent inwardly (a); P 1 exp terminal seta straight (p)
22. Labrum plough-like (a); Labrum not transformed nor prominent (p)
23. A 2 exp absent (a); A 2 exp present (p)
24. P 5 in total with 7 setae (a); P 5 in total with 6 setae (p)
25. P 3 enp remains 3-seg; P 2, 4 enp 2-seg (a); P 2–4 enp 3-seg (p)
26. A 2 exp with 2 setae (a); A 2 exp with 1 seta (p)
27. P 2–4 enp absent (a); P 2–4 enp 3-seg (p)
28. P 2–4 exp III with 3 setae (a); P 2–4 exp III with more than 3 setae (p)
29. P 1 exp setae long (a); P 1 exp setae stout (p)
30. Anal operculum with large denticles (a); anal operculum with small denticles (p)
31. A 2 exp absent (a); A 2 exp present (p)
32. Mxp basis with 2 setae (a); Mxp basis with 1 seta (p)
33. Habitus vermiform (a); habitus fusiform (p)
34. Cu. r. setae short (a), Cu. r. setae long (p)

7 setae is otherwise only present in *M. gorbunovi*, *M. smirnovi*, *M. atlantica* and *M. beckeri* within *Metahuntemannia*, thus not in the species under consideration.

Considering a 3-segmented state of the P 2–4 endopodite as plesiomorphic, the P 3 endopodite remaining 3-segmented but always the P 2 and P 4 becoming 2-segmented (25) is a peculiar autapomorphy of a group comprising *M. smirnovi*, *M. triarticulata* and *M. dovpori*. An antennal exopodite carrying 2 setae (26) is the only case within *Metahuntemannia*, all the other species bear 1 seta.

The phylogenetic value of the absence of the P 2–4 enp (27) is arbitrary for the taxa under consideration (see discussion above) and is proposed to be a weak autapomorphy here. Both sistertaxa, *M. drzycimskii* and *M. texturata* share three synapomorphies of which the 3 setae on P 2–4 exp III (28) is the strongest for it does not occur in any other representative of *Metahuntemannia*. Compared to other species of the group under consideration longer setae of the P 1 exp and larger denticles on the anal operculum also represent an apomorphic state. However, it was not possible to compare the latter character with *M. atlantica* and *M. arctica* for this character is not yet described for these species.

Unfortunately, we could not find any autapomorphy for *M. gorbunovi*, *M. atlantica*, *M. arctica* and *M. beckeri*. However, the spinulated lobe on the inner margin of the caudal rami is remarkably prominent in these species. Spinules and possibly secretion products of the dorsal portion of the caudal rami are the cause for detritic particles to be glued to the inner portion of the rami hampering a careful investigation. This phenomenon has otherwise only been observed at *M. crassa* by KLIE (1941) and possibly provide another autapomorphy supporting the monophyly of a group of species consisting of: *M. smirnovi*, *M. texturata*, *M. gorbunovi*, *M. atlantica*, *M. arctica*, *M. beckeri*, *M. spinosa* and *M. indica*. *M. arctica* bears most apomorphic characters. Although all three autapomorphies (i. e. P 1 exp III with 3 setae, Md enp fused, A 2 allobasis with 1 seta) occur in other species of *Metahuntemannia* as well (see discussion above), it is a new acquisition for *M. arctica*, compared to its closest relatives. As for *M. gorbunovi*, *M. atlantica* and *M. beckeri* character differences are small, either due to real similarity or to the lack of detail in the descriptions available. For these species, diagnostic characters are described in order to differentiate between them (see key above). However, these characters allow no conclusion on the phylogenetic relationships of their semaphoronts yet.

The presence of 2 setae on the Mxp basis (32) is a remarkable autapomorphy for *M. spinosa* for to our knowledge it is not known from any other taxon of Cletodidae. Another peculiar character of *M. spinosa* has to be mentioned here: it is the single seta on the prospective endopodite of P 5.

The so far monotypic *Beckeria* Por, 1986 is of remarkable resemblance to all known species of *Metahuntemannia* and especially to *M. spinosa*. Characters of striking similarity not listed in Figs. 37 and 38 include: principle morphology and peculiarities of A1, Md, Mx 1, Mx 2 and Mxp. As in *Metahuntemannia* the anal operculum is armed with denticles at its caudal margin. The only characters differing are the vermiform habitus (33) and the caudal rami (shape, reduced seta number and reduced length of the major terminal seta) (34) shown in the description of *Beckeria indica*. The vermiform habitus is estimated as being less important for this may also be due to a fixation artefact. Transformations of caudal rami have taken place in various ways among Harpacticoida, however, a reduction of setal size (34) to this extent is not known so far from any representative of Cletodidae. POR (1986 a: 83) admitted that "*Beckeria* could be situated among the species of *Metahuntemannia*" but established a new taxon with the "broad and extremely modified furcal branches" as its sole constitutive character. Therefore, we suspect the validity of this taxon and synonymize it herein with *Metahuntemannia*.

As for *Talpina* (Fig. 38) two groups can safely be distinguished: the one with 1-segmented P 1 exopodites in the females (47) and mandibular endopodites comprising of 3 setae (48) with the bulk of 8 species (namely *T. peruana, T. pectinata, T. pacifica, T. micracantha, T. furcispina, T. noodti, T. fodens* and *T. bathyalis*). The second group comprising of *T. bifida, T. curticauda* and *T. talpa* with 3-segmented P 1 exopodites in the females and 1–2 setae on the mandibular endopodites as the plesiomorphic state is characterized by the autapomorphy: P 1 enp 1-segmented with a thorn-like element (40) which is not present among other species nor in other taxa of Huntemanniidae Por, 1986 (a single thorn without segment (7), however, is constitutive for *Metahuntemannia*). As discussed above for *Metahuntemannia* is the reduction of P 2–4 endopodites (41) a common phenomenon among the taxa under consideration; therefore, its phylogenetic significance is restricted.

T. bifida and *T. curticauda* share 4 setae on the P 1 exp III (42), a peculiarity not known from other representatives of *Metahuntemannia* nor *Talpina* (but 4 setae are present in e. g. *Huntemannia*). A similarly strong autapomorphy is the reported absence of the P 5 (43) otherwise not known from Huntemanniidae Por, 1986.

An armature comprising only of 1 seta on the mandibular endopodite (45) in *T. bifida* is the lowest reported for *Talpina*: in other species it is 2 or mostly 3 (cf. 48).

A P 2 endopodite remnant (46) as in *T. curticauda* is otherwise not known from Huntemanniidae Por, 1986.

T. peruana and *T. pacifica* share a peculiar triangular plate representing the fused P 1 exopodites of the females (49). For *T. pectinata* this is assumed for no female of this species is known till now. These three species share the male peculiarity within *Metahuntemannia* and *Talpina* of a separate P 5 exp (51). *T. pacifica* is characterized by a 2-segmented P 2 endopodite (52) which, however, is assumed to have become 2-segmented in *T. pectinata* independently.

T. peruana is characterized by the fusion of even the basis and the endopodite of P 1 (53) – a character of a leg except P 5 to our knowledge not known from any other harpacticoid.

T. pectinata bears a dagger-shaped cutting edge of the mandible (54) and rows

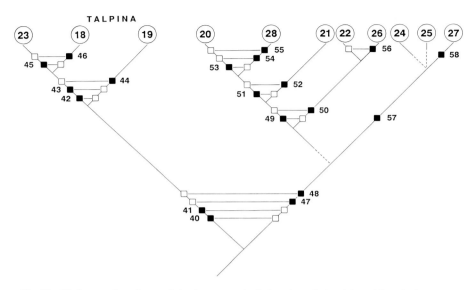

Fig. 38. Cladogram based on adult characters depicting the relationships within *Talpina* (a = apomorphic; p = plesiomorphic). Character states as follows (encircled numbers refer to the species as listed in Table 1; sole numbers refer to characters):
40. P 1 enp 1-seg with thorn-like element (a); P 1 enp 1-seg with 1 plumose seta (p)
41. P 2–4 enp absent (a); P 2–4 enp at least 2-seg (p)
42. P 1 III with 4 setae (a); P 1 III with 3 setae (p)
43. P 5 lacking in the females (a); P 5 present in both sexes (p)
44. A 2 exp with 2 setae (a); A 2 exp with 3 setae (p)
45. Md enp with 1 seta (a); Md enp with 2 setae (p)
46. P 2 enp indicated by a small protuberance (a); P 2 enp absent (p)
47. P 1 exp 1-seg in the females (a); P 1 exp 3-seg in the females (p)
48. Md enp with 3 setae (a); Md enp with 1–2 setae (p)
49. P 1 exp a triangular plate in the females (unknown for *T. pectinata*) (a); P 1 exp not triangular (p)
50. Mx 2 syncoxa elongate (a); Mx 2 syncoxa short, endites close together (p)
51. P 5 exp separate in the males (a); P 5 exp fused with benp (p)
52. P 2 enp 2-seg (a); P 2 enp 3-seg (p)
53. P 1 basis and enp fused (a); P 1 basis and enp separate (p)
54. Md cutting edge dagger-shaped (a); Md cutting edge of complex architecture (p)
55. Ultimate abdominal segment with peculiar ventral spinule pattern (a); spinule pattern absent (a)
56. Caudal ramus with inner spiniform outgrowth (a); caudal ramus without inner outgrowth (p)
57. P 2–4 enp absent (a); P 2–4 enp present (p)
58. A 2 exp absent (a); A 2 exp present (p)

of large spinules on the ventral side of the abdomen (55). These are convincing autapomorphies for they are not known from any other representative of Huntemanniidae Por, 1986.

T. micracantha and *T. furcispina* share an elongate syncoxa of the maxilla (50), providing a huge gap between the proximal and the distal endites. This character

Table 1: Sampling site characteristics of all representatives belonging to *Metahuntemannia* and *Talpina* gen. nov. (f = female, m = male, cop. = copepodid; ff/mm = more than one female/male)

Species	Number	Length (μm)	Locality	Date	Depth (m)	Reference
1. *M. spinosa*	1f	700	Iceland	7/1896	980	KLIE 1941
2. *M. gorbunovi*	2f/1 cop. f	710	Arctic Ocean	1936/37		SMIRNOV 1946
3. *M. crassa*	1f	650	Raunefjorden near Bergen (Atlantic)	12/1962	112	POR 1965
4. *M. smirnovi* (*M. drzycimskii* nom. nov.)	4f	620–650	Coast off Bergen (Atlantic)	4/1967	300	DRZYCIMSKI 1968 SOYER 1970
5. *M. smirnovi*	1m	450	Golf de Gascogne		2050	BODIN 1968
6. *M. doveori*	2f	430	Golf de Gascogne		3950	BODIN 1968
7. *M. mediterranea*	2f	520/550	Mediterranean Sea		106	SOYER 1970
8. *M. magniceps*	1f	610	Peru Trench	11/1965	2000	BECKER et al. 1979
9. *M. iberica*	1f	385	Iberian Deep Sea	3/1970	3820	BECKER et al. 1979
10. *M. pseudomagniceps*	1f/2m	485f/250m	Iceland-Faroer-Ridge	7/1966	464	SCHRIEVER 1983
11. *M. atlantica*	2f	670	Iceland-Faroer-Ridge	7/1966	500	SCHRIEVER 1983
12. *M. triarticulata*	2f/2m	560f/480m	Iceland-Faroer-Ridge	7/1966	1540	SCHRIEVER 1984
13. *M. arctica*	1f	650f	Iceland-Faroer-Ridge	7/1966	1825	SCHRIEVER 1984
14. *M. texturata*	1f	875	Weddell Sea	10/1986	250	present study
15. *M. beckeri*	1f	825	Weddell Sea	11/1986	390	present study
16. *M. spinipes*	1f	740	Weddell Sea	12/1986	270	present study
17. *M. indica*	4f		Indian Ocean	9/1964	450	POR 1986
18. *T. curticauda*	1f	470	Iberian Deep Sea	3/1970	4000	BECKER et al. 1979
19. *T. talpa*	ff/2m	660f/345m	Peru Trench	10/1965	920	BECKER et al. 1979
20. *T. peruana*	2f/mm	580f/420m	Peru Trench	10/1965f 11/1965m	4100f 2000m	BECKER et al. 1979
21. *T. pacifica*	1f/1m	504f	Peru Trench	10/1965	920	BECKER et al. 1979
22. *T. micracantha*	1f	2200	Mariana Trench	8/1980	5750	GAMO 1981
23. *T. bifida*	2f/2cop.	500	Iceland-Faroer-Ridge	7/1966	985	SCHRIEVER 1984
24. *T. noodti*	1f	1530	Weddell Sea	11/1986	430	present study
25. *T. fodens*	1f	1400	Weddell Sea	11/1986	400	present study
26. *T. furcispina* sp. n.	1f	1630	Weddell Sea	2/1989	2000	present study
27. *T. bathyalis*	1f	790	Weddell Sea	4/1990	3970	present study
28. *T. pectinata*	1m	1150	Weddell Sea	2/1989	400	present study

state is otherwise not present in *Metahuntemannia* and *Talpina*; it is merely indicated in *T. noodti*. *T. micracantha* and *T. furcispina* share also a remarkably large body size for both species are the largest representatives of *Metahuntemannia* and *Talpina* (cf. Table 1). However, there are differences: as for the antennule (in *T. furcispina* there are no plumose and stout setae but 5 instead of 4 setae on the lateral margin of the distal segment and 2 instead of 1 seta accompanying the aesthetasc; penultimate segment with only 1 instead of 5 armature elements); antenna (no allobasis but distinct basis in *T. furcispina*, middle seta of endopodite stouter; both the outermost setae not plumose but ornamented with spinules); mandibular endopodite with the second most largest seta at the inner corner in *T. furcispina* but only a small remnant in *T. micracantha*; most strikingly different are the endopodites of P 2–3, 3-segmented and with different armature in *T. micracantha* and 2-segmented in *T. furcispina*.

An inner spiniform outgrowth of the caudal ramus (56) of *T. furcispina* is not known from any other representative of Huntemanniidae Por, 1986. Therefore, it provides a convincing autapomorphy of this species (such an outgrowth is described from *Huntemannia* but in this taxon arising from the outer edge of the caudal rami).

As discussed above the presence or absence of a P 2–4 enp is not a very convincing character for its reduction (57) is assumed to have taken place for more than once independently (at least once in *Metahuntemannia* and twice in *Talpina*). Unfortunately, we have not found an autapomorphy for neither *T. noodti* nor for *T. fodens*.

The absence of the antennal exopodite (58) in *T. bathyalis* is the only case within *Talpina*, in the other species it is 1-segmented bearing 1–3 setae.

Distribution

The 20 species of *Metahuntemannia* and *Talpina* hitherto known to science are reported from all over the worlds oceans: as from the Arctic and Antarctic portion of the Atlantic, the Iceland, Norwegian, Biscayan and Northwest African waters of the Atlantic, the Mediterranean and the Mariana and Peru Trench of the Pacific (Fig. 39). However, these taxa are restricted to deeper waters than 100 m, most of its representatives collected in the deep sea down to a depth of 6300 m (the latter reported for *T. talpa* by BECKER (1979) from the Peru Trench) (Table 1). The shallowest occurrence is reported from 106 m depth of the Mediterranean for *M. mediterranea* (SOYER 1970) and from 112 m depth of the Raunefjorden (Atlantic) for *M. crassa* (POR 1965).

The high Antarctic representatives of *Metahuntemannia* and *Talpina* have been found all over the sampling area from most shallow waters at the shelf ice

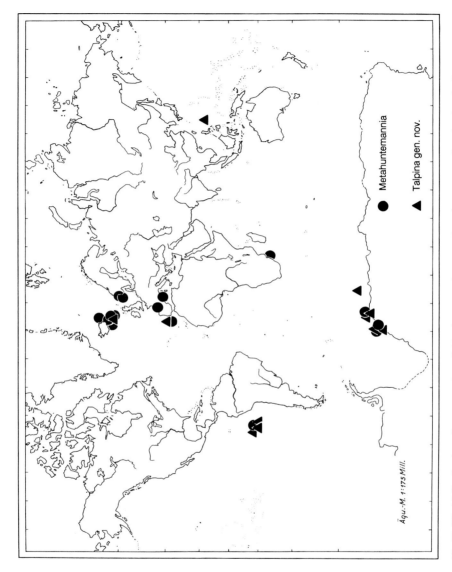

Fig. 39. World distribution of representatives of *Metahuntemannia* (except the unknown sampling site given by SMIRNOV 1946 from the Arctic Ocean) and *Talpina*.

edge of 270 m depth to 3970 m depth (deepest station investigated) of the Weddell Sea basin. The sampling stations of the new Antarctic species range from about 66° to 74° S and from about 25° W to 34° E (Fig. 1).

It has to be emphasized that representatives of *Metahuntemannia* and *Talpina* are very sparsely distributed: not more than about 41 females and 11 males belonging to the 28 species reported from the worlds oceans have been found (including the 8 new species described herein). Sixteen of these species are described on the basis of one specimen only. In 22 species only one sex is known, in 20 species purely the female. Additionally, in those cases where more than one specimen per species were found, this was either in the same sample or the same geographical area; e. g. *T. talpa*, *T. peruana* and *T. pacifica* described by BECKER (1979). As for *T. peruana* there is a discrepancy between its record given in the thesis of BECKER where a male is reported from 5000 m depth and BECKER et al. (1979) where this information is lacking (cf. Table 1). Actually, none of the 28 species of both taxa known today was found in any other topographic unit of the ocean floor suggesting on a remarkable degree of endemism on species level. High diversity and a large portion of endemic taxa is commonly believed to be characteristic for the deep-sea environment (e. g. SANDERS 1979).

Cletodidae (LANG 1948, cf. POR 1986 b) is commonly known to be a harpacticoid family typical for the deep sea and muddy habitats. However, there are examples of representatives occurring in muddy substrates also of shallow waters, even in the Wadden Sea of the German Bight, North Sea (e. g. *Enhydrosoma propinquum*, *E. longifurcatum*, *Rhizothrix (R.) minuta*, *Nannopus palustris* i. a. cf. DAHMS 1985).

Life habits

Representatives of *Metahuntemannia* and especially of *Talpina* are well adapted to an inbenthic mode of life as evidenced by a cylindrical body shape, stout appendages, spiniform setae on the antennules, reductions in the locomotory leg endopodites, and setal lengths of the exopodites. Females of *Talpina* are even better adapted to a burrowing life style than those of *Metahuntemannia* due to their vermiform body shape (the *spinosa*-group with a fusiform habitus), plough-like rostrum, P 1 without intercoxal plate but fused coxae, stouter setae and exopodal segments than in *Metahuntemannia*. Especially the P 1 is specialized in *Talpina* for its digging function as indicated by an elongate coxa, a medially flexed exopodite and stout exopodal segments armed with spiniform setae forming a shovel together with the reduced endopodite. The exopodites of the first swimming legs are flexed here medially almost to a right angle in the females. Subterminal and lateral tube pores and an apical spinule on the spiniform setae of the P 1 exopodite indicate additional functions (secretory, sensory?) of the first locomotor appendages (Fig. 40). Functional investigations of living representatives of *Metahuntemannia* are hampered by their deep sea occurrence. Instead, behavioural studies of the less extremely specialized taxon *Huntemannia*, easy to obtain from the shallow North Sea (KUNZ 1971), are suggested.

Specialized appendages for digging and burrowing are known from animals of various groups. The transformation of the frontalmost legs used for burrowing became a well known example for convergent evolution. This occurs independently in different invertebrate and vertebrate groups, for instance in insects (e. g. *Gryllotalpa vulgaris*/ Saltatoria; *Lyristes plebejus*/ Rhynchota; *Arenivaga investigata*/ Blattodea; *Scarites buparius*/ Coleoptera), chelicerates (*Siloannea macroceras*/ Araneida), tetrapod reptiles (e. g. *Palmatogecko rangii*/ Squamata) or mammals (e. g. *Tachyglossus a. aculeatus*/ Monotremata, *Notoryctes typhlops*/ Marsupialia; *Talpa europaea*/ Insectivora; *Cricetus vulgaris*/ Rodentia).

In the aquatic meiofauna, however, specialized digging appendages are so far only known from *Metahuntemannia*. As all the species of this taxon are found in muddy sediments BECKER in his thesis hypothesized the first locomotor appendages to be "Grabschaufeln" for burrowing into the fluffy substrate. Detailed description of this specialized appendages with considerations on their functional implications have so far been lacking.

As for the feeding habits some species of *Metahuntemannia* and *Talpina* at least are supposed to be predatory or scavenging for the fact that the cutting edge of the mandibles is tapering and drawn out as being two-tipped in *M. iberica*, *T. curticauda* and *T. talpa* (these species being described by BECKER 1979). The most remarkable mandible in this respect is that of *T. pectinata* male described in the present study which is tapering to a single point. Such mandibles must be suitable for seizing prey, carcasses or tissue. Feeding on nematodes or fish-larvae is reported from *Tisbe furcata* (MARCOTTE 1977), and DAHMS et al. (1991) show the remarkable elongation of the *corpus mandibulae* of this species. Predatory feeding habits are otherwise hypothesized for deep-sea Harpacticoida by KUNZ (1984).

As reported above the specimens of *Metahuntemannia* known so far comprise of 41 females and 11 males at least (in the study of BECKER (1979) no precise numbers for the females and males of *T. talpa* and *T. peruana* are mentioned). The ratio of about 4 females to 1 male is even outnumbered by the present study where we found 7 adult females belonging to *Metahuntemannia* and *Talpina* but only 1 male, namely of *Talpina pectinata* sp. n.

The sex ratio of copepods in the deep sea has been the subject of several studies (e. g. THISTLE & ECKMANN 1990). The results reported therein also provide highly female biassed sex ratios.

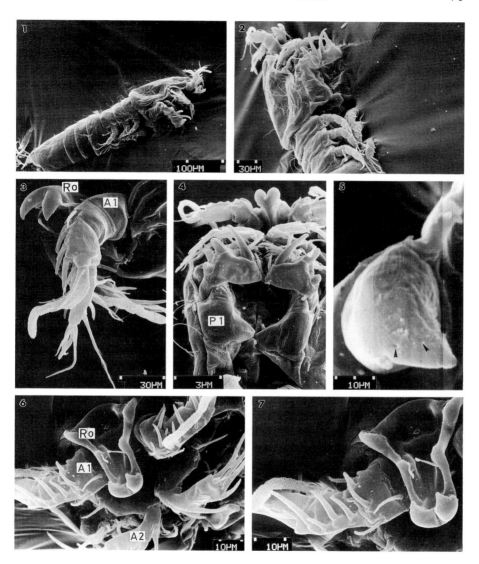

Fig. 40. *Talpina furcispina* sp. n., scanning electron micrographs of copepodid IV. 1–2. Habitus in ventrolateral view. 3. Plough-like rostrum and right antennule. 4. Cephalothorax in ventral view with elongate P 1 in posterior view always flexed frontad. 5. Outer lateral spine of right P 1 exopodite with lateral pores (arrowheads). 6–7. Cephalothorax in frontal view (A 1 = antennule; A 2 = antenna; Ro = rostrum; P 1 = first locomotor thoracopod).

As for the reproductive biology of the taxa under consideration, it is surprising that neither in the at least 34 adult females reported in the literature nor in the 7 females studied here, eggs or egg sacs have been found. This could be explained by the ceasing of reproductive activity at the times of sampling. However, sampling has taken place at different times of the year, also of the specimens investigated in the present study. Therefore, it is argued that females of *Metahuntemannia* carry their eggs either only for a short period of time as it is known from a representative of Parastenocarididae (GLATZEL 1990) or deposit the eggs immediately after protrusion. This behaviour is also reported for Chappuisiidae (LANG 1948) and is proposed for Neobradyidae and Metidae for the reason that eggs have never been observed to be carried by the females (DAHMS unpubl. observ.).

In this context we may mention the observation of i. a. POPPE (1884) that the related *Huntemannia* occurring in the shallow marine waters, is characterized by another peculiarity of the females. These females always bear 2 egg sacs, a phenomenon among Harpacticoida otherwise only known from the Canuellidae (Polyarthra) and Diosaccidae (LANG 1948), not considering the pair of eggs carried only for a short time by *Parastenocaris phyllura* (GLATZEL 1990).

Zusammenfassung

Das Taxon *Metahuntemannia* Smirnov, 1946 wird in die beiden Schwestergruppen *Metahuntemannia* und *Talpina* aufgeteilt. Zu *Metahuntemannia* s. str. gehören die Vertreter der von BECKER (1979) unterschiedenen *spinosa*-Gruppe, zu *Talpina* gen. n. die später beschriebenen Vertreter der *talpa*-Gruppe. Neben den bereits bekannten werden zusätzliche Autapomorphien der beiden Taxa beschrieben und ein Bestimmungsschlüssel bereitgestellt. *Beckeria* Por, 1986 wird mit *Metahuntemannia* synonymisiert. Von Meiobenthosproben dreier Expeditionen des Fs POLARSTERN zur hochantarktischen Weddell See werden drei neue Arten von *Metahuntemannia*, nämlich *M. texturata*, *M. beckeri* und *M. spinipes*, sowie fünf neue Arten von *Talpina* beschrieben: *T. noodti*, *T. fodens*, *T. furcispina*, *T. bathyalis* und *T. pectinata*. Diese Funde lassen auf eine erhebliche Diversität und einen hohen Grad an Endemismus dieser Gruppe in der Hochantarktis schließen. Die Tiere weisen eine endopelische Lebensweise auf und zeigen folgende Anpassungen: eine zylindrische Gestalt, gestauchte Extremitäten, dornenförmige Bewehrungen, Reduktionen der Schwimmbeinendopoditen und der Borstenlänge der Exopodite. Vertreter von *Talpina* weisen einen P 1 auf, der durch eine verlängerte Coxa und einen mediad gebogenen Exopoditen zu einer „Grabschaufel" umgewandelt ist. Die Weibchen beider

Taxa tragen ihre Eier offensichtlich nicht lange am Körper. Das Geschlechterverhältnis ist wie bei anderen Tiefseecopepoden erheblich zugunsten der Weibchen verschoben. Zusammen mit den bereits bekannten 13 Arten von *Metahuntemannia* und 6 Arten von *Talpina* werden die verwandtschaftlichen Verhältnisse und die Zoogeographie der 28 Arten diskutiert.

Acknowledgements

We are indebted to Prof. Dr. H. K. Schminke (Oldenburg) for his generosity, encouragement, counsel and providing us with laboratory facilities. We want to thank Prof. Dr. P. Ax, Dr. W. Mielke (Göttingen) as well as Dr. G. Schriever (Kiel) for valuable criticism on the manuscript. Prof. Schminke and the "Alfred-Wegener-Institute for Polar and Marine Research" (Bremerhaven) offered H.–U. D. to participate in the Ant V/3 and Ant VII/4 expeditions of RV Polarstern to the Weddell Sea (Antarctica). Dr. R. L. Herman (Gent) as coordinator of the EPOS meiobenthos-group is acknowledged for providing us with additional material from the EPOS cruise. Data presented here were also taken during the European "Polarstern" study (EPOS) sponsored by the European Science Foundation and the "Alfred-Wegener-Institute for Polar and Marine Research". Mrs. B. Hosfeld is acknowledged for collecting meiofauna during the Ant VIII/6 cruise. Mrs. H. Bünker helped largely in preparing the typed version of the text. Research support was provided through the "Deutsche Forschungsgemeinschaft".

References

BECKER, K. H. (1979) (W. NOODT & G. SCHRIEVER Hrsg.): Eidonomie und Taxonomie abyssaler Harpacticoidea (Crustacea, Copepoda). II. Paramesochridae, Cylindropsyllidae und Cletodidae. "Meteor" Forsch.-Ergebn., Dtsch. **31**, 1–37.

BODIN, P. (1968): Copépodes Harpacticoides des étages bathyal et abyssal du Golfe de Gascogne. Mem. Mus. Nat. Hist. nat., Sér. A, **55** (1), 1–107.

DAHMS, H.-U. (1985): Zur Harpacticidenfauna der Jade (innerer Teil) im Tidebereich. Drosera **85** (2), 65–76.

DAHMS, H.-U.; SCHMINKE, H. K. & M. POTTEK (1991): A redescription of *Tisbe furcata* (Baird, 1837) (Copepoda, Harpacticoida, Tisbidae) and its phylogenetic relationships within the taxon *Tisbe*. Z. zool. Syst. Evolut.-forsch. **29**: 433–449.

DRZYCIMSKI, I. (1968): *Metahuntemannia* Smirnov und *Apodella* Por (Cop. Harp.) mit Beschreibung einer neuen Art aus dem westnorwegischen Küstengebiet. Sarsia **31**, 127–130.

GAMO, S. (1981): *Metahuntemannia micracantha* sp. nov., a new abyssal harpacticoid copepod (Crustacea) from east of the Mariana Trench. Bull. Biogeogr. Soc. Japan **36** (1), 1–7.

GLATZEL, T. (1990): On the biology of *Parastenocaris phyllura* Kiefer (Copepoda, Harpacticoida). Stygologia **5**, 131–136.

KLIE, W. (1941): Marine Harpacticoiden von Island. Kiel. Meeresforsch. 5, 1–44.
KUNZ, H. (1971): Verzeichnis der marinen und Brackwasser bewohnenden Harpacticiden der deutschen Meeresküste. Kieler Meeresforsch. 27 (1), 73–93.
LANG, K. (1948): Monographie der Harpacticiden I und II. Otto Koeltz Science Publishers, Koenigstein, W-Germany, 1–1682.
MIELKE, W. (1984): Some remarks on the mandible of the Harpacticoida (Copepoda). Crustaceana 46, 257–260.
POPPE, S. A. (1884). Die freilebenden Copepoden des Jadebusens. Abh. nat. Ver. Bremen 9, 167–206.
POR, F. D. (1965): Harpacticoida (Crustacea, Copepoda) from muddy bottoms near Bergen. Sarsia 21, 1–16.
POR, F. D. (1986 a): New deepsea Harpacticoidea (Copepoda) of cletodid type, collected in the Indian Ocean of R/V "Anton Bruun" in 1964. Crustaceana 50 (1), 78–98.
POR, F. D. (1986 b): A re-evaluation of the family Cletodidae Sars, Lang (Copepoda, Harpacticoida). Syllogeus 58, 420–425.
SCHRIEVER, G. (1983): New Harpacticoidea (Crustacea, Copepoda) from the North Atlantic Ocean. III. New species of the family Cletodidae. "Meteor" Forsch.-Ergeb., Dtsch. 36, 65–83.
SCHRIEVER, G. (1984): New Harpacticoidea (Crustacea, Copepoda) from the North Atlantic Ocean. V. Three new species of *Metahuntemannia* Smirnov (Cletodidae). Zool. Scr. 13 (4), 277–284.
SMIRNOV, C. C. (1946): New species of Copepoda Harpacticoida from the Arctic Ocean. Trudy Exped. Glavsevmov. Ledokol. Par. "Sedov" 3, 231–263.
SOYER, J. (1970): Contribution à l'étude des Copépodes Harpacticoides de Méditerranée Occidentale. 3. Découverte du genre *Metahuntemannia* Smirnov. Vie Milieu 21 (2–A), 279–286.

Dr. Hans-Uwe Dahms and *Mark Pottek*
Fachbereich 7 (Biologie), Arbeitsgruppe Zoomorphologie, Universität Oldenburg
Postfach 25 03, D-2900 Oldenburg

Description of some benthic Copepoda from Chile and a discussion on the relationships of *Paraschizopera* and *Schizopera* (Diosaccidae)

Wolfgang Mielke

Contents

Abstract	79
A. Introduction	80
B. Results	80
Additional findings: *Tisbe* spec., *Amphiascus* spec., *Nitocra* spec.	80
Idomene Philippi, 1843	80
Idomene cookensi Pallares, 1975	80
Schizopera Sars, 1905	82
Schizopera chiloensis nov. spec.	82
Phyllopodopsyllus T. Scott, 1906	92
Phyllopodopsyllus mossmani T. Scott, 1912 *chiloensis* nov. subspec.	92
Zusammenfassung	98
Acknowledgement	99
References	99

Abstract

Several benthic copepod species from Maiquillahue and Quellón (Chile) are presented. *Tisbe* spec., *Amphiascus* spec., and *Nitocra* spec. were determined only up to genus level. Until now *Idomene cookensi* Pallares, 1975 was only known from the Argentine Isla de los Estados. Only recently some animals of this species have been found in Playa Maiquillahue. *Schizopera chiloensis* and *Phyllopodopsyllus mossmani chiloensis* are new findings. The only habitat of both species known so far is the beach of Quellón Viejo situated south-east of the island of Chiloé.

The genus *Eoschizopera*, established by WELLS & RAO (1976), with which the specimens of *Schizopera chiloensis* could be classed without problems, is rejected, because it is based only on plesiomorphic features.

A. Introduction

This report presents some copepod species, which were collected in the central/southern part of the Chilian Pacific coast in 1983 (MIELKE 1985). It deals with species from coarse sediments of Maiquillahue, north of Valdivia, and Quellón, situated at the south-eastern coast of the island of Chiloé. A number of copepods of these localities were already described in the publication cited above. Now, some additional information is given, including the descriptions of a new subspecies and a new species.

B. Results

Additional findings: Some species were determined only up to genus level:

Tisbidae: *Tisbe* spec. (Playa Maiquillahue, 28 February 1983. Corresponds locality 4 a in MIELKE 1985; 35 ♀ ♀ – 21 of them ovigerous –, 8 ♂ ♂.

Diosaccidae: *Amphiascus* spec., *varians*-group (Playa Maiquillahue, 28 February 1983. Corresponds locality 4 a; 11 ♀ ♀, 7 ♂ ♂. Quellón Viejo, 8 March 1983. Corresponds locality 5; 4 ♀ ♀ – 3 of them ovigerous –, 7 ♂ ♂. The animals of both localities seemingly belong to one species).

Ameiridae: *Nitocra* spec. (Isla Maiquillahue, 28 February 1983. Corresponds locality 4 b; 95 ♀ ♀ – 40 of them ovigerous –, 19 ♂ ♂. Quellón Viejo, 8 March 1983. Corresponds locality 5; 6 ♀ ♀ – 2 of them ovigerous –, 2 ♂ ♂.

In Playa Maiquillahue (28 February 1983; locality 4 a) one representative of the Thalestridae was found which can be identified without any problems with *Idomene cookensi* Pallares, 1975 (figs. 1 A–E). Two ♀ ♀ (0,43 and 0,44 mm) and 1 ♂ (0,45 mm) were collected. Apart from some minor differences they show only one appreciable divergence from the animals of the Isla de los Estados (PALLARES 1975): Between both apical setae on distal segment of endopodite P 2 ♂ another rather thin appendage is to be seen (fig. 1 C). Since only 1 ♂ from Maiquillahue was available, it cannot be decided whether this arrangement is exceptional or characteristic of the total population.

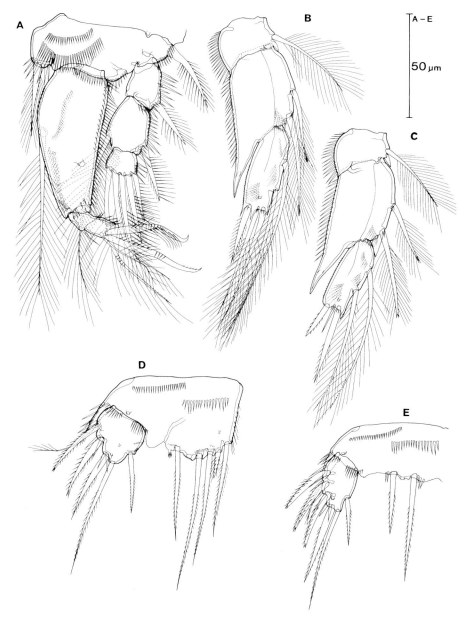

Fig. 1. *Idomene cookensi*. A. P 1 ♀. B. Enp. P 2 ♀. C. Enp. P 2 ♂. D. P 5 ♀. E. P 5 ♂.

Diosaccidae Sars, 1906
Schizopera Sars, 1905
Schizopera chiloensis nov. spec.

(Figs. 2–6)

Locality and material. Quellón Viejo, south-eastern coast of Isla Chiloé (Locus typicus; 8 March 1983. Corresponds locality 5 in MIELKE 1985); 21 ♀♀, 13 ♂♂.
12 animals were dissected (6 ♀♀, 6 ♂♂). Holotype female, reg. no. I Chi 344; 4 paratypes (2 ♀♀, 2 ♂♂), reg. no. I Chi 341, 342, 343, 350. With the exception of habitus, abdomen ventral, rostrum, mandible, and labium all drawings of the female are from the holotype.

Description

Female: Body length from tip of rostrum to end of furcal rami 0,47–0,53 mm (holotype 0,51 mm). Rostrum (fig. 3 A) slender, apically prolonged to a nipple and with two hairlike setae laterally. Dorsal caudal part of cephalothorax with some (five) tongue-like lappets. With the exception of the last somite, which bears spinules ventrally, the distal margins of all somites are bare. Subdistally – with the exception of ultimate and penultimate somites – all somites are furnished with hairlike setae. Genital double-somite only with a short sign of subdivision laterally. Genital area see figs. 2 B, C. Distal margin of penultimate somite extended dorsally to a half-round plate (pseudoperculum). Anal operculum seemingly bare, a row of fine setules is to be seen beneath it. Furcal rami scarcely tapering caudally. Outer edge with a slender seta, accompanied by a rather short appendage. Dorsal and ventral surface each with a slender seta; apical edge with three setae, innermost one short, central one longest (figs. 2 A–C).

1. Antenna (fig. 3 B): 8 segments. Fourth and last segment each with an aesthetasc.
2. Antenna (fig. 3 C): Basis and endopodite with an indicated sign of subdivision. First enp. segment with one seta. Second enp. segment with several spinules, two strong and two hairlike setae subapically; distal edge with seven appendages altogether. Exopodite 3-segmented. Basal and short middle segment with one seta each, distal segment with one seta laterally, two strong and apparently one fine accompanying setae distally.

Mandible (fig. 3 D): Chewing edge of precoxa with several teeth and a plumose seta. Coxa-basis with rows of hairs and spinules and three plumose setae. Endopodite subapically with two, apically with five setae, three of which are fused basally. Exopodite small, with two slender setae (or one which is bifurcated?).

Labium: See fig. 3 E.

1. Maxilla (fig. 3 F): Edge of precoxal arthrite with eight (nine?) appendages;

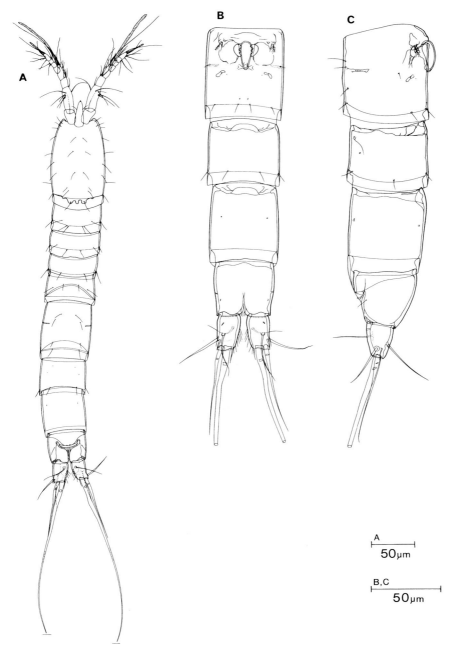

Fig. 2. *Schizopera chiloensis* nov. spec. ♀. A. Habitus, dorsal side. B. Abdomen, ventral side. C. Abdomen, lateral side.

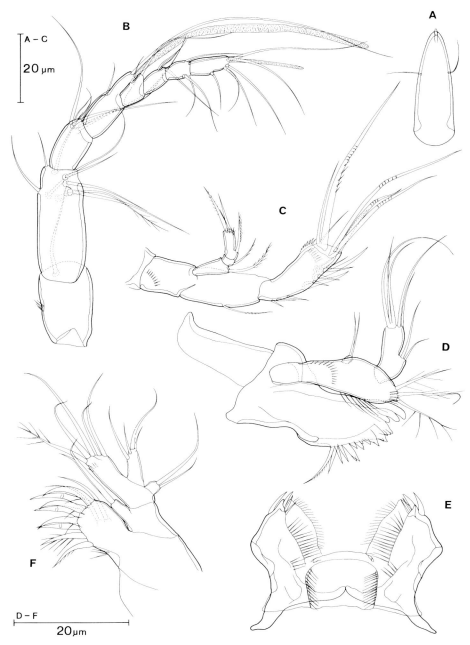

Fig. 3. *Schizopera chiloensis* nov. spec. ♀. A. Rostrum. B. 1. Antenna. C. 2. Antenna. D. Mandible. E. Labium. F. 1. Maxilla.

dorsal surface with two setae. Coxa, basis, endopodite and exopodite with 2, 5, 4, 2 setae, respectively.

2. Maxilla (fig. 4 A): Syncoxa with three endites having 3, 2, 3 plumose setae, respectively. Basis with one claw and two setae. Endopodite probably two-segmented with six setae altogether.

Maxilliped (fig. 4 B): Basis with slender spinules and three plumose setae. Proximal endopodite-segment with long setules and two setae on inner edge. Distal endopodite-segment furnished with one claw and three setae.

P 1 (fig. 4 C): Coxa and basis with several rows of spinules; basis with one outer and one inner strong seta. Endopodite three-segmented. Proximal segment prolonged, more or less of the same length as the whole exopodite. Inner margin subapically with a slender seta. Middle segment short, furnished only with some spinules. Distal segment with two slender claw-like appendages and one accompanying short seta. Exopodite three-segmented. Proximal segment longest, distal segment with four appendages.

P 2–P 4 (figs. 4 E; 5 A,B): Coxa with a few rows of spinules. Basis with one outer seta, the one of P 3 longest. Endopodites three-segmented. Proximal segment spinulose on inner and outer edge, that of P 4 with one inner seta. Distal outer edge of middle segment prolonged to a tooth-like projection, with one inner seta. Distal segment longest, also acutely prolonged distally, with 4, 3, 2 appendages, respectively. Exopodites three-segmented. Proximal and middle segments with a denticulate hyaline frill distally, middle segment with an inner seta. Distal segment longest, with four (P 2 and P 3) and five (P 4) appendages altogether. Distal outer edge of all three segments acutely prolonged.

Seta and spine formula:		Exopodite	Endopodite
	P 2	(0.1.022)	(0.1.121)
	P 3	(0.1.022)	(0.1.111)
	P 4	(0.1.122)	(1.1.011)

P 5 (figs. 5 C,D): Baseoendopodite with a long outer seta; inner part with four appendages. Exopodite with five or six slender setae.

Male: Differing from the female in the following respects:
– Body length 0,44–0,49 mm.
– 1. Antenna haplocer.
– Inner part of basis P 1 modified (fig. 4 D).
– Enp. P 2 shows sexual dimorphism (fig. 4 F).
– Last segment exp. P 3 with an inner hyaline appendage (fig. 5 E).
– Baseoendopodites of both P 5 fused. Each benp. with a long outer seta and

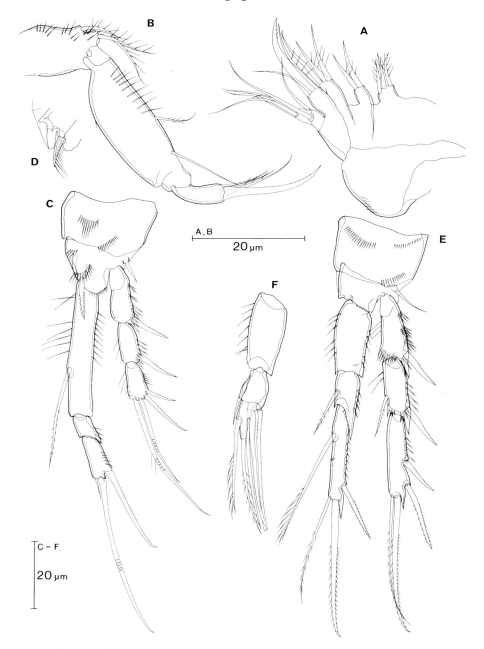

Fig. 4. *Schizopera chiloensis* nov. spec. A. 2. Maxilla ♀. B. Maxilliped ♀. C. P 1 ♀. D. Inner part of basis P 1 ♂. E. P 2 ♀. F. Enp. P 2 ♂.

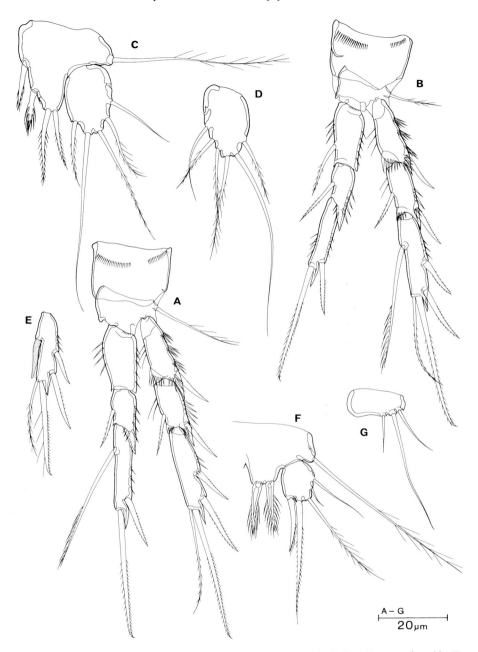

Fig. 5. *Schizopera chiloensis* nov. spec. A. P 3 ♀. B. P 4 ♀. C. P 5 ♀. D. Exp. P 5 ♀, other side. E. Terminal segment of exp. P 3 ♂. F. P 5 ♂. G. P 6 ♂.

two strongly plumose setae on inner part. Exp. with five setae of different length (fig. 5 F).
– P 6 forming a transverse plate bearing three setae, middle one longest (fig. 5 G).

Variability. The setation of P 5 ♀ is rather variable. The exopodite can be furnished with five or six setae, respectively. There are specimens with five setae on one P 5, and six setae on the other one of the same pair (figs. 5 C,D). Occasionally the inner part of benp. bears only three setae.

Discussion. The genus *Schizopera* Sars, 1905 has repeatedly been subject of discussions on its ecology and phylogenetic relationships. On the one hand it represents a group, the species of which occur in freshwater as well as in brackish and marine water, raising the question about the original milieu and probable ways of invasion (e. g. LANG 1948, CHAPPUIS 1953, NOODT & PURASJOKI 1953). On the other hand the species share some features by which they can obviously be delimited from other groups without problems. Since the publication of LANG's monograph (1948), especially WELLS & RAO (1976) and APOSTOLOV (1982) each gave a detailed paper on the extent and relationships of the genus. Resulting from these papers was a subdivision of the *Schizopera* species into three genera comprising two subgenera, respectively:

Eoschizopera Wells & Rao, 1976
    Eoschizopera s. str.
    Praeschizopera Apostolov, 1982
Schizopera Sars, 1905
    Schizopera s. str.
    Neoschizopera Apostolov, 1982
Schizoperopsis Apostolov, 1982
    Schizoperopsis s. str.
    Psammoschizoperopsis Apostolov, 1982

Furthermore, two genera were erected apparently belonging to the group of taxa closely related to *Schizopera*: *Paraschizopera* Wells, 1981 (= "*Paraschizopera* sp." in BECKER & SCHRIEVER 1979) and (questionably) *Schizoperoides* Por, 1968. Until now *Paraschizopera* has only been known as a female copepodite (designated as *P. beckeri* by WELLS 1981). Likewise, in *Schizoperoides*, the male has not been described until today (POR 1968).

The specimens of Quellón Viejo, Chiloé can be easily classed with *Eoschizopera*, a genus established in 1976 by WELLS & RAO. In this paper, the authors united all *Schizopera* species showing conspicuous plesiomorphic features (2. antenna with basis, setation P 2–P 4. Also the male setae bearing P 6 may be mentioned here, which seems to exist only in *Eoschizopera*). Furthermore, some features were given, which indeed can be interpreted as apomorphic, but apomorphic at a higher level of the systematic hierarchy (Ax 1987), i. e. which hold true at least for the entire *Schizopera* group. An autapomorphy proving the existence of a monophylum *Eoschizopera* was not given (I could not find any likewise). In all probability, it represents a paraphylum based on symplesiomorphies, that means an artificial group of species which must be rejected.

Summarizing the actual knowledge of the features (mouth parts and genital area are known only insufficiently to some extent and are therefore not considered) of the *Schizopera* relationgroup the following diagram of the phylogenetic relationship can be given (fig. 6). The assumption that *Paraschizopera* represents the adelphotaxon of *Schizopera* surely is questionable, above all, because the male of *Paraschizopera* is not known. Nevertheless, the representation can be regarded as a legitimate working hypothesis. It must be emphasized that nearly all given apomorphies are negative features. Merely the existence of the hyaline spine on the distal segment exp. P 3 ♂ has to be interpreted as an apomorphic positive feature. Above all LANG (1948, 1965) already pointed out this phylogenetically very valuable structure. WELLS & RAO (1976), too, underlined its significance and "universality" and confirmed its presence in several species, for which this information had not been given by its first authors. Recently GEE & FLEEGER (1990) have investigated a number of representatives of Diosaccidae with regard to the distal segment of exp. P 3 ♂. They found that the hyaline spine of the *Schizopera* relationgroup (*Schizopera*, *Schizoperopsis* and *Eoschizopera*) represents an "articulating appendage" in contrast to the hyaline tubes also existing on that segment in other genera.

The difficulties concerning features of reduction are evident. Since the reduction of setae, spines or segments is a frequent phenomenon within the Copepoda "Harpacticoida", convergences have to be taken into account (concerning *Schizopera* see NOODT & PURASJOKI 1953 and LANG 1965). The decisive question, whether two or more taxa represent a monophylum or an artificial paraphylum cannot be answered without the knowledge of features which undoubtedly represent synapomorphies. Nevertheless, it must be emphasized that a feature of reduction which evidently has developed convergently in two or more taxa may absolutely be a valuable autapomorphy for each of these taxa. For example, the reduction to two outer spines on the distal segment exp. P 2–P 4 has taken place several times within the Diosaccidae. The conclusion of a closer relationship because of this convergence is inadmissible. However, this feature is of great importance for, e. g., the taxon *Schizopera* (Group B alone or even group A).

The diagram (see fig. 6 and table 1) suggests the following inferences:
(1) At least in exp. of 2. antenna and enp. P 1 parallel reductions of segments have taken place.

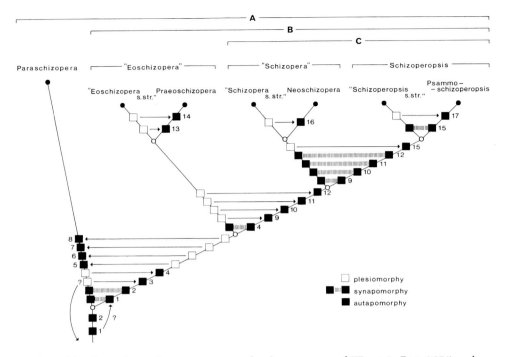

Fig. 6. The figure shows the attempt to transfer the statements of WELLS & RAO (1976) and APOSTOLOV (1982) on the systematics of the *Schizopera* relationgroup into a diagram of the phylogenetic relationship. As a result it can be demonstrated that "*Eoschizopera*", "*Eoschizopera* s. str." and "*Schizoperopsis* s. str." are no monophyla because no autapomorphy for each of these groups could be established. On the other hand the groups A–C and *Schizoperopsis* apparently represent monophyletic taxa. To the purpose of this study it is not necessary to designate categorial terms to the groups A and C. However, the taxon B is equated with the genus *Schizopera* and most probably represents the adelphotaxon of *Paraschizopera*.

(2) It is still unsure, whether or not the groups A–C and *Schizoperopsis* really represent monophyla as the scheme alleges (groups basing on features of reduction; questionable systematical position of *Paraschizopera*). At least the taxon B has to be interpreted as a monophylum provided that the existence of the hyaline spine on distal segment exp. P 3 ♂ is restricted to this group.

(3) If not even one autapomorphy exists for "*Eoschizopera*" Wells & Rao, 1976 as well as for "*Schizopera*" sensu Apostolov, 1982, then, they are only paraphyla which do not represent equivalents to unities in living Nature (AX 1987).

Unmasked paraphyla are to be rejected remorselessly. The division of the *Schizopera* species by APOSTOLOV 1982 is only of diagnostic value. For me all

Table 1 (compare fig. 6. Plesiomorphies in angular parentheses).

(1) Female genital field (according to LANG 1965, WELLS & RAO 1976).
(2) Distal segments of exp. P 2–P 4 only with 2 outer spines [3].
(3) Distal segment of exp. P 3 ♂ with a hyaline spine [without]. The situation of *Paraschizopera* is unclear because the male is still unknown.
(4) Distal segment of exp. P 4 with at most 1 inner seta [2].
(5) 2. Antenna with allobasis [basis].
(6) Exp. 2. antenna only with 1 segment [3]. In *S. anomala* Coull, 1971 and *S. paradoxa* Daday, 1904 the exp. 2. antenna is also 1-segmented. This must be interpreted as a convergence.
(7) Enp. P 2–P 4 with 2 segments [3].
(8) The state of P 5 ♀ is rather questionable. Probably benp. and exp. are fused in adult *Paraschizopera* (see BECKER & SCHRIEVER 1979) and originally subdivided in *Schizopera* (group B). But there exist a number of *Schizopera* species having a at least partly P 5 ♀ with fused benp. and exp. Since these species do not represent a monophyletic taxon, the fusion must have taken place convergently.
(9) Distal segment of exp. P 4 without inner seta [1]. The reduction of the inner seta took place in the stem line of group C. *Eoschizopera crassispinata*, too, has no inner seta. But seemingly CHAPPUIS (1954) has confounded P 3 with P 4 (?). So *E. marlieri* (see ROUCH & CHAPPUIS 1960) seems to be the only *Eoschizopera* species which has no inner seta; probably a convergent feature.
(10) 2. Antenna with allobasis [basis].
(11) Exp. 2. antenna with 2 segments [3].
(12) Reduction of P 6 ♂ [present]. The presence in the *Eoschizopera* species *crassispinata*, *marlieri* and *gligici* is unknown.
(13) Exp. 2. antenna with 2 segments [3].
(14) Enp. P 1 with 2 segments [3].
(15) Enp. P 4 with 2 segments [3].
(16) Enp. P 1 with 2 segments [3].
(17) Enp. P 1 with 2 segments [3].

species of the monophyletic group B belong to the taxon *Schizopera* given the categorial rank of a genus. But according to AX (1987) "all categorial terms applied to taxa above species taxa are nothing but arbitrary labels". Practical considerations, i. e., decrease in species number or narrowing of diagnoses must not be a justification for the erection of artificial groups.

The new species *Schizopera chiloensis* differs from all "*Eoschizopera*" species in seta and spine formula.

To my knowledge the following *Schizopera* species are known from the neotropical region: *S. gauldi* Chappuis & Rouch, 1961 from Brazil (ROUCH 1962); *S. giselae* Alvarez, 1988 from Brazil (ALVAREZ 1988); *S. haitiana* Kiefer, 1934 from Haiti (KIEFER 1934); *S. noodti* Rouch, 1962 from Argentina (ROUCH 1962); *S. pori* Alvarez, 1988 from Brazil (ALVAREZ 1988); *S. tobae* Chappuis, 1931 and *S. tobae cubana* Petkovski, 1973 from Cuba (PETKOVSKI 1973); *S. triacantha* Kiefer, 1934 from Haiti (KIEFER 1934); *S. vicina* Herbst, 1960 from Peru (HERBST 1960) and Brazil (REID & ESTEVES 1984).

The genus *Schizopera* was mentioned but not determined up to species level by MARINONI (1964) from Brazil and by EBERT & NOODT (1975) from Chile.

Tetragonicipitidae Lang, 1948
Phyllopodopsyllus T. Scott, 1906
Phyllopodopsyllus mossmani T. Scott, 1912
chiloensis nov. subspec.

(Figs. 7–10)

Locality and material. Quellón Viejo, south-eastern coast of Isla Chiloé (Locus typicus; 8 March 1983. Corresponds locality 5 in MIELKE 1985); 4 ♀♀ (1 ovigerous), 5 ♂♂ and 1 copepodite.
5 animals were dissected (3 ♀♀, 2 ♂♂). Holotype female, reg. no. I Chi 367; 4 paratypes (2 ♀♀, 2 ♂♂), reg. no. I Chi 366, 368, 369 and II Chi 13 a-m. 1. antenna, 1. and 2. maxilla, maxilliped, P 1, P 2, P 4, P 5 are drawn from holotype.

Description
Female: Body length from tip of rostrum to end of furcal rami 0,52–0,58 mm (holotype 0,58 mm). Rostrum short, defined at base; subapically with two fine setae (fig. 7 A). Distal margins of cephalothorax and pereiomeres bare. Distal margin of genital double-somite and following somite with spinules laterally and dorsally. Penultimate somite with spinules round about. Anal somite seemingly bare. Genital double-somite subdivided. Genital area see fig. 9 B. Anal operculum with spinules. Furcal rami irregularly shaped, laterally flattened. Outer edge with two slender setae, subapically on dorsal side a seta inserts which is biarticulated at its base. The long apical seta is irregularly broadened basally, fused with a small seta and accompanied by a plumose seta. Apical margin with spinules (fig. 9 B).

1. Antenna (fig. 7 A): 9 segments, first of which is the longest one. Second segment without toothlike projection. Fourth and last segment each with an aesthetasc.

2. Antenna (fig. 7 B): Basis with some fine spinules. Exopodite one-segmented, with three plumose appendages, outermost is basally fused with exp. First endopodite-segment only with some spinules. Second endopodite-segment with rows of spinules; subapically with 3, apically with 7 appendages altogether.

Mandible (fig. 7 C): Chewing edge of precoxa with several teeth and a plumose seta. Coxa-basis with 3 plumose setae and two rows of long spinules on surface. Endopodite with 2 setae subapically and 7 setae apically. Exopodite bearing 2 setae.

1. Maxilla (fig. 7 D): Edge of precoxal arthrite with about 11 appendages;

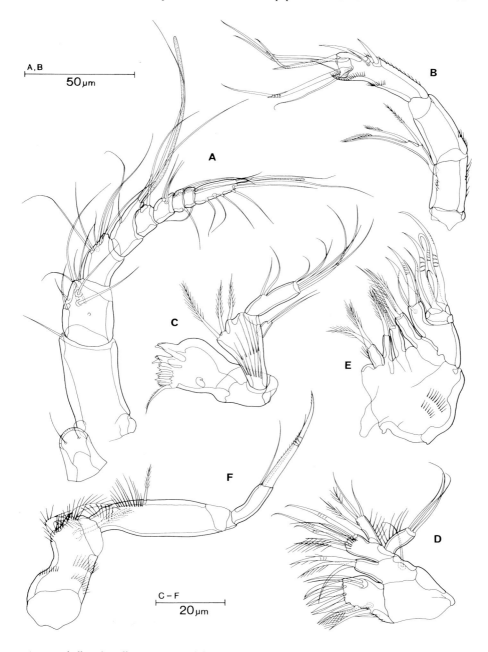

Fig. 7. *Phyllopodopsyllus mossmani chiloensis* nov. subspec. ♀. A. Rostrum and 1. Antenna. B. 2. Antenna. C. Mandible. D. 1. Maxilla. E. 2. Maxilla. F. Maxilliped.

dorsal surface with 2 setae. Coxa having 3 setae on endite and 1 seta on exite. Basis, endopodite and exopodite are furnished with 7, 4, 3 setae, respectively.

2. Maxilla (fig. 7 E): Syncoxa with four endites bearing 2, 1, 3, 3 setae, respectively. Basis with 2 strong appendages and 2 slender setae. Endopodite seemingly two-segmented, with 5 appendages altogether.

Maxilliped (fig. 7 F): Basis with several rows of spinules and 3 plumose setae. Proximal endopodite-segment with a row of long hairlike spinules and 1 seta on inner edge. Distal endopodite-segment with 1 claw and 2 slender setae.

P 1 (fig. 8 A): Coxa with rows of spinules. Basis armed with spinules and a strong appendage on inner and outer edge. Exopodite three-segmented. Middle segment with long spinules on inner edge. Distal segment with 4 setae. Endopodite two-segmented. Basal segment prolonged, inner edge with a row of hairs and a plumose seta subapically. Distal segment short with rows of spinules and two strong appendages of different length.

P 2 – P 4 (figs. 8 B, C; 10 A): Coxa with rows of spinules on surface. Outer appendage of basis strong in P 2 and slender in P 3 and P 4. Exopodites three-segmented. Proximal segment with an inner seta. Proximal and middle segment with a toothlike projection on distal outer edge. Terminal segment of P 2 and P 3 also with a toothlike prolongation apically. The segment has 4 appendages in P 2 and P 3, and 7 appendages in P 4. Endopodites two-segmented. Proximal segment with an inner seta. Distal segment acutely prolonged, with 3 setae, innermost is the longest one.

Seta and spine formula:

	Exopodite	Endopodite
P 2	(1.0.112)	(1.021)
P 3	(1.0.112)	(1.021)
P 4	(1.0.322)	(1.021)

P 5 (fig. 9 A): Outer and distal edge with 7, inner edge with 4 setae of different length.

M a l e : Differing from the female in the following respects:
- Body length 0,45–0,48 mm.
- With the exception of the first abdominal somite all other somites are furnished ventrally with spinules on distal margin.
- 1. Antenna subchirocer.
- Enp. P 2 two-segmented. Proximal segment with an inner seta. Distal segment with 3 appendages, inner one shortest, outer one longest and threadlike tapering (fig. 8 D).
- Enp. P 3 compared to that one of female with different proportions in length (fig. 8 E).

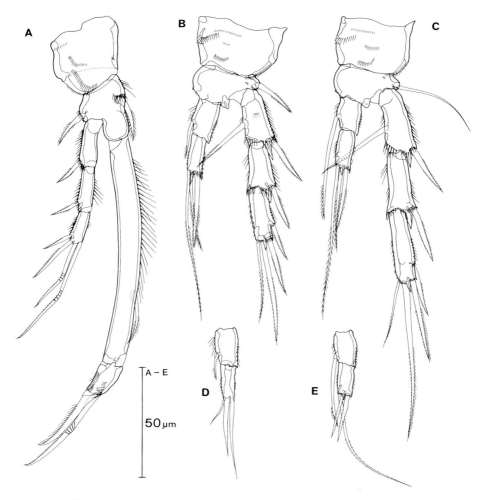

Fig. 8. *Phyllopodopsyllus mossmani chiloensis* nov. subspec. A. P 1 ♀. B. P 2 ♀. C. P 3 ♀. D. Enp. P 2 ♂. E. Enp. P 3 ♂.

- Distal segment of exopodite P 4 only with 6 setae. Distal segment of endopodite only with two appendages, inner of which is strong and curved (fig. 10 B).
- P 5 of both sides fused. Baseoendopodite with 1 long outer seta and 3 setae on inner part. Exopodite bearing 4 setae, innermost one longest (fig. 10 C).
- P 6 with two setae, inner one plumose (fig. 10 D).
- Furcal rami tapering, about twice as long as broad. Outer edge with two

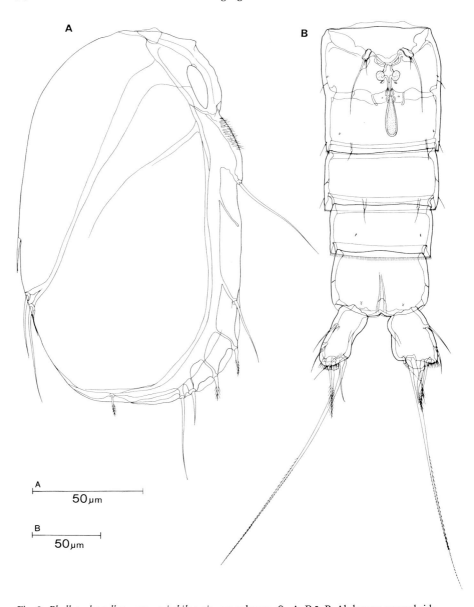

Fig. 9. *Phyllopodopsyllus mossmani chiloensis* nov. subspec. ♀. A. P 5. B. Abdomen, ventral side.

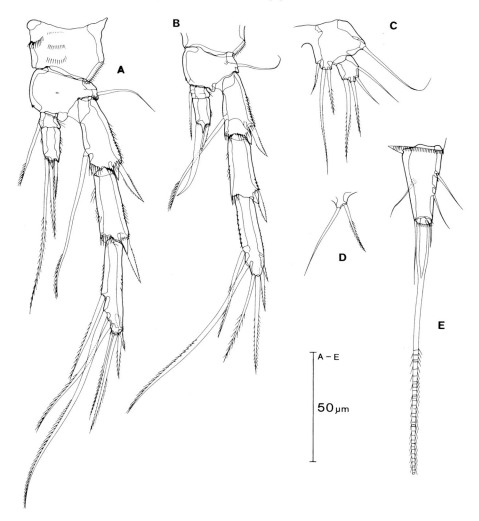

Fig. 10. *Phyllopodopsyllus mossmani chiloensis* nov. subspec. A. P 4 ♀. B. P 4 ♂. C. P 5 ♂. D. P 6 ♂. E. Furca, ventral side ♂.

setae; dorsal seta inserting in the center. Apical seta thickened basally, distal part plumose. Outside a small seta is fused with the apical seta, another slender inner seta is accompanying it (fig. 10 E).

Discussion. The animals from Isla Chiloé show two important conformities, a 9-segmented, toothless 1. antenna and the identical seta and spine formula of P 1 – P 4, with some other *Phyllopodopsyllus* species:

P. berrieri Monard, 1936. The ♀ was reported from Algiers, North Africa (MONARD 1936). The present animals differ from *P. berrieri* in the structure of the furca and in the proportions in length of exp. and enp. P 1. PESTA (1959) found the ♂ in the Gulf of Naples. He stated among other things that the exp. P 5 bears 5 setae (♂ from Chiloé only with 4 setae).

P. hibernicus (Roe, 1955). Unfortunately only the ♀ of this Irish species is known (ROE 1955). ROE stresses her specimens to have a "slight protuberance on the second segment" of the 1. antenna. Other small differences are the proportions in length of exp. and enp. P 1 and the structure of the furca.

P. laspalmensis Marinov, 1973. The species was reported from Las Palmas, Canary Islands. Only 1 ♀ was available, the ♂ missing like in the preceding species. The present individuals do not exhibit clear deviations from MARINOV's (1973) description; probably the furca shows some slight differences. On the other hand *P. laspalmensis* is not fundamentally different from *P. mossmani*. However, MARINOV only compared his animal with *P. thiebaudi* Petkovski, 1955.

P. mossmani T. Scott, 1912. The species was collected by T. SCOTT (1912) from Port Stanley, Falkland Islands (Islas Malvinas). Later, PALLARES (1982) redescribed the species from Bahía Thetis, Argentine part of Tierra del Fuego and Pto. Argentino, Islas Malvinas. The animals from Chiloé largely correspond with this description. There are only some minor differences concerning proportions, furcal structure and body length. The only crucial point seems to be P 5 ♂. I could only observe 4 setae on exp. instead of 5 as drawn by T. SCOTT and PALLARES. Therefore I establish a new subspecies for the animals from Isla Chiloé: *Phyllopodopsyllus mossmani chiloensis*.

Zusammenfassung

Von Maiquillahue und Quellón (Chile) werden einige benthale Copepodenarten vorgestellt. *Tisbe* spec., *Amphiascus* spec. und *Nitocra* spec. wurden lediglich bis zur Gattung bestimmt.

Idomene cookensi Pallares, 1975 war bisher nur von der argentinischen Isla de los Estados bekannt. Einige Exemplare dieser Art fanden sich nun auch im Strand von Maiquillahue.

Schizopera chiloensis und *Phyllopodopsyllus mossmani chiloensis* sind neu für die Wissenschaft. Einziger Fundort dieser beiden Formen ist bis jetzt der Strand von Quellón Viejo im Südosten der Insel Chiloé.

Die von WELLS & RAO (1976) aufgestellte Gattung *Eoschizopera*, zu der die Individuen von *Schizopera chiloensis* problemos zu stellen wären, wird zurückgewiesen, da sie lediglich auf plesiomorphen Merkmalen basiert.

Acknowledgement

The research trip was made possible by a grant of the Deutsche Forschungsgemeinschaft (Mi 218/2–1).

References

ALVAREZ, M. P. J. (1988): Harpacticoid copepods from Una do Prelado River (São Paulo, Brazil): genus *Schizopera*. Hydrobiologia **167/168**, 435–444.

APOSTOLOV, A. M. (1982): Genres et sous-genres nouveaux de la famille Diosaccidae Sars et Cylindropsyllidae Sars, Lang (Copepoda, Harpacticoidea). Acta Zool. Bulg. **19**, 37–42.

AX, P. (1987): The Phylogenetic System. The Systematization of Organisms on the Basis of their Phylogenesis. John Wiley & Sons, Chichester etc., 340 pp.

BECKER, K.-H. & G. SCHRIEVER (1979): Eidonomie und Taxonomie abyssaler Harpacticoidea (Crustacea, Copepoda). Teil III. 13 neue Tiefsee-Copepoda Harpacticoidea der Familien Canuellidae, Cerviniidae, Tisbidae, Thalestridae, Diosaccidae und Ameiridae. "Meteor" Forsch.-Ergebnisse **31**, 38–62.

CHAPPUIS, P. A. (1953): Harpacticides psammiques récoltés par Cl. Delamare Deboutteville en Méditerranée. Vie Milieu **4**, 254–276.

– (1954): Recherches sur la faune interstitielle des sédiments marins et d'eau douce à Madagascar. IV. – Copépodes Harpacticoides psammiques de Madagascar. Mém. Inst. sci. Madagascar **9**, 45–73.

EBERT, S. & W. NOODT (1975): Canthocamptidae aus Limnopsammon in Chile (Copepoda Harpacticoidea). Gew. Abwässer **57/58**, 121–140.

GEE, J. M. & J. W. FLEEGER (1990): *Haloschizopera apprisea*, a New Species of Harpacticoid Copepod from Alaska, and Some Observations on Sexual Dimorphism in the Family Diosaccidae. Trans. Am. Microsc. Soc. **109**, 282–299.

HERBST H. V. (1960): Copepoden (Crustacea, Entomostraca) aus Nicaragua und Südperu. Gew. Abwässer **27**, 27–54.

KIEFER, F. (1934): Neue Ruderfußkrebse von der Insel Haiti. Zool. Anz. **108**, 227–233.

LANG, K. (1948): Monographie der Harpacticiden. Nordiska Bokhandeln, Stockholm, 1682 pp.

– (1965): Copepoda Harpacticoidea from the Californian Pacific coast. Kungl. Svenska Vetenskaps. Handl. **10**, 1–566.

MARINONI, R. C. (1964): *Diarthrodes falcipes* n. sp. (Harpacticoidea – Copepoda) encontrada em algas do litoral catarinense. Bol. Univ. Paraná **2**, 59–73.

MARINOV, T. M. (1973): *Phyllopodopsyllus laspalmensis* n. sp. und *Eurycletodes aberrans* n. sp. aus dem Atlantischen Ozean. C. r. Acad. bulg. Sci. **26**, 1525–1528.

MIELKE, W. (1985): Interstitielle Copepoda aus dem zentralen Landesteil von Chile: Cylindropsyllidae, Laophontidae, Ancorabolidae. Microfauna Marina **2**, 181–270.

MONARD, A. (1936): Note préliminaire sur la faune des Harpacticoides marins d'Alger. Bull. Trav. publ. Sta. Aquic. Pêche Castiglione 1–41.

NOODT, W. & K. J. PURASJOKI (1953): *Schizopera ornata* n. sp., ein neuer Copepode aus Brackwasserbiotopen der deutschen und finnischen Ostseeküste. Soc. Sci. Fenn. Comm. Biol. **13**, 1–10.

PALLARES, R. E. (1975): Copépodos harpacticoides marinos de Tierra del Fuego (Argentina). I. Isla de los Estados. Contr. Cient. CIBIMA **122**, 1–34.

– (1982): Copépodos harpacticoides marinos de Tierra del Fuego (Argentina). IV. Bahía Thetis. Contr. Cient. CIBIMA **186**, 1–52.

PESTA, O. (1959): Harpacticoiden (Crust. Copepoda) aus submarinen Höhlen und den benachbarten Litoralbezirken am Kap von Sorrent (Neapel). Pubbl. Staz. Zool. Napoli **30**, 95–177.

PETKOVSKI, T. K. (1973): Subterrane Süßwasser-Harpacticoida von Kuba. (Vorläufige Mitteilung.) Rés. exp. biosp. cub. roum. Cuba, Acad. R. S. Rom., Bucuresti **1**, 125–141.

Por, F. D. (1968): Copepods of some land-locked basins on the islands of Entedebir and Nocra (Dahlak Archipelago, Red Sea). Sea Fish. Res. St. Haifa Bull. **49**, 32–50.

Reid, J. W. & F. De Esteves (1984): Considerações ecológicas e biogeográficas sobre a fauna de copépodos (Crustacea) planctónicos e bentónicos de 14 lagoas costeiras do estado do Rio de Janeiro, Brasil. Restingas: Origem, Estrutura, Processos. Lacerda, L. D. de et al. (orgs.), CEUFF, Niterói, Brazil. p. 305–326.

Roe, K. M. (1955): Genus *Paraphyllopodopsyllus* Lang (Copepoda Harp.). Two new species – *P. hibernicus* and *P. hardingi.* Proc. Roy. Ir. Acad. **57**, 131–139.

Rouch, R. (1962): Harpacticoides (Crustacés Copépodes) d'Amérique du Sud. Biol. Amér. Austr. **1**, 237–280.

Rouch, R. & P. A. Chappuis (1960): Sur quelques Copépodes Harpacticoides du lac Tanganika. Rev. Zool. Bot. Afr. **61**, 283–286.

Scott, T. (1912): The Entomostraca of the Scottish National Antarctic Expedition, 1902–1904. Trans. Roy. Soc. Edinb. **48**, 521–599.

Wells, J. B. J. (1981): Keys to aid in the identification of marine harpacticoid copepods. Amendment Bulletin No. 3. Zool. Publ. Vict. Univ. Wellington **75**, 1–13.

Wells, J. B. J. & G. C. Rao (1976): The relationship of the genus *Schizopera* Sars within the family Diosaccidae (Copepoda: Harpacticoida). Zool. Journ. Linn. Soc. **58**, 79–90.

Dr. Wolfgang Mielke
II. Zoologisches Institut und Museum der Universität Göttingen
Berliner Straße 28, D-3400 Göttingen

Six representatives of the Tetragonicipitidae (Copepoda) from Costa Rica

Wolfgang Mielke

Contents

Abstract	101
A. Introduction	102
B. Localities, material	102
C. Results	107
Tetragonicipitidae Lang, 1948	107
Laophontella Thompson & A. Scott, 1903	107
Laophontella horrida (Por, 1964) *dentata* nov. subspec.	107
Oniscopsis Chappuis, 1954	118
Oniscopsis robinsoni Chappuis & Delamare Deboutteville, 1956	118
Phyllopodopsyllus T. Scott, 1906	120
Phyllopodopsyllus setouchiensis Kitazima, 1981	120
Phyllopodopsyllus ancylus nov. spec.	126
Phyllopodopsyllus carinatus nov. spec.	133
Phyllopodopsyllus gertrudi Kunz, 1984 *costaricensis* nov. subspec.	141
Zusammenfassung	144
Acknowledgements	144
References	145

Abstract

Six representatives of the Tetragonicipitidae are recorded from different beaches of Costa Rica.

Laophontella horrida dentata is established as a new subspecies.

The individuals of two further species are identified with *Oniscopsis robinsoni* Chappuis & Delamare Deboutteville, 1956 and *Phyllopodopsyllus setouchiensis* Kitazima, 1981.

P. ancylus and *P. carinatus* are described as new species.

The new subspecies *P. gertrudi costaricensis* is compared to the nominate subspecies *P. gertrudi gertrudi* Kunz, 1984.

A. Introduction

Until now the marine coastal fauna of Costa Rica has merely been investigated in part. The knowledge of the macrofauna is obviously more comprehensive than of the microfauna. A number of macrofaunal publications is available. Vicariously the records on polychaetes (MAURER et al. 1988) and isopods (BRUSCA & IVERSON 1985) or the paper on the macrofaunal community structure by VARGAS (1987) should be mentioned here.

Recently, also the investigation of the marine meiofauna was initiated (see VARGAS 1988; some more references are given in this paper) dealing with ecological subjects on the meiofaunal major taxa of the Punta Morales (Gulf of Nicoya) area.

To my knowledge no informations on species level of marine benthic copepods have been given anywhere. On the other hand some data exist on freshwater copepods (e. g. COLLADO et al. 1984 a, b; DUSSART & FERNANDO 1985 a, b; LÖFFLER 1972; REID 1990) mostly treating planktonic species.

During my stay in Costa Rica (23. 8.–20. 9. 1990) I had the opportunity to collect a great number of benthic copepods from different beaches of the Caribbean and the Pacific coast. Main object of the trip was (1) to give a preliminary inventory of the unknown marine benthic copepod fauna and (2) to trace possible zoogeographical relations between Costa Rican copepods and species of the Caribbean and Pacific area.

B. Localities, material

The Central American country Costa Rica covers about 51 000 km^2 and extends between a latitude of 08° 02' – 11° 13' N and a longitude of 82° 34' – 85° 58' W. The North of Costa Rica is bordered by Nicaragua and its South adjoins Panamá (fig. 1). The Atlantic (Caribbean) coastline comprises some more than 200 km mainly consisting of sandy beaches. The Pacific coast stretches over a length of about 1200 km and offers more diversity than the Atlantic side. Numerous beaches, bays etc., often framed by rocky formations, exhibit sediments of different material and grain size. At the Pacific side the tidal range can reach about 3 m, whereas at the Caribbean coast the difference between low and high tide line usually is less than 50 cm.

The Pacific side of Costa Rica zoogeographically belongs to the tropical Panamanian (Panamic) Province, a part of the Eastern Pacific Region which extends between Punta Eugenia/Baja California at 28° N (according to BRUSCA

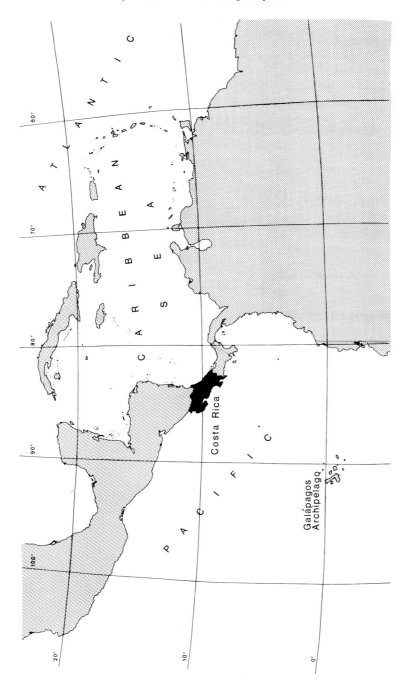

Fig. 1. Geographical position of Costa Rica.

& IVERSON 1985) or the southern tip of Baja California at 23° N (see LAGUNA 1990) and the Gulf of Guayaquil at 3° S.

The localities are briefly characterized in the following (see also figs. 2 and 3):

Caribbean coast:

1) Playa Portete, near Limon (6 September 1990): Exposed beach of small width and about 60–80 m in length. Substratum rather coarse.

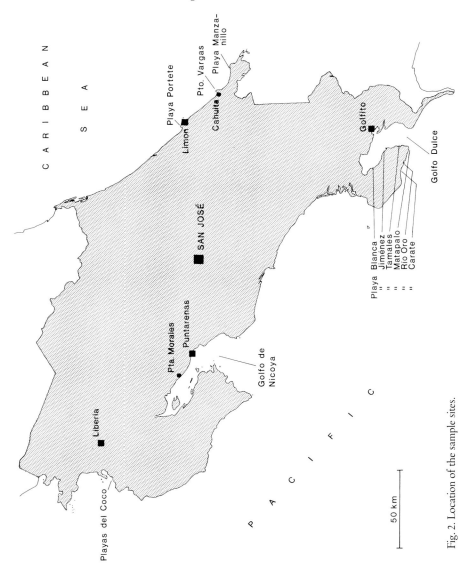

Fig. 2. Location of the sample sites.

Fig. 3. A. Beach of Punta Morales, Gulf of Nicoya, Pacific side. B. Beach near Manzanillo, Caribbean coast.

2) Puerto Vargas, near Cahuita (7 September 1990): Very long beach, grain size middle or fine.

3) Playa Manzanillo, east of Pto. Viejo (7 September 1990): Beach segment behind the village. Light and relatively coarse sediment.

Pacific coast:

4) Punta Morales, a small peninsula on the east coast of the Gulf of Nicoya (27–29 August 1990): At Punta Morales there are mangrove swamps as well as sandy beaches isolated by rocky sections. The "main" beach is about 30 m broad and a few 100 m long. White and coarse sediment containing much detritus, because large amounts of whirled up particles are outwelled into the Gulf by several rivers during the rainy season (May – November). Mean tidal range 2,3 m. Salinity fluctuating according to rainfalls between 25‰ and 34‰ (mostly about 32‰). Surface water temperature above 25° C (see VARGAS 1987, 1988; VOORHIS et al. 1983).

Several samples were taken and, in addition, two fragmentary transects. On that occasion the samples were taken from different distances of the slope based on the low tide line. Thus, most species can be recorded which in principle have each a definite specific distribution pattern.

5) Punta Morales (30 August 1990): (a) Samples from small pools about upper eulittoral/supralittoral. Bottom of pools grown with green algae. (b) Copepods from sponges in rockpools.

6) Playas del Coco near Liberia in the north-west of Costa Rica (1 September 1990): Some samples of a fragmentary transect. Fine dark sand, little detritus.

Península Osa (9 and 10 September 1990): A few samples, respectively, from four exposed (7–10) and two sheltered (11–12) beaches.

7) Playa Carate at the south of the peninsula. Km-long beach. Fine sand near surface, deeper layers with gravel and stones.

8) Playa Rio Oro. Heavy waves. Steep slope.

9) Playa Matapalo. Corresponds to the previous beach.

10) Playa Tamales. Situated at the entrance of the Golfo Dulce; can still be characterized as exposed beach. Km-long beach of dark substratum.

11) Playa Jiménez. Situated within the Golfo Dulce. Inner bay of the small town. Amount of detritus clearly higher than in the previous beaches.

12) Playa Blanca. Sheltered beach. Fine and coarse sand together. Much detritus.

13) Enlarged city beach of Puntarenas, Gulf of Nicoya (17 September 1990). Km-long beach, sheltered type. Fine, brownish sand.

The samples yielded a few thousand benthic copepods. At each locality cited above the sediment was put into plastic bags and usually transported to the laboratory of the CIMAR in Punta Morales as soon as possible. There, the sediment was given into a small container and marine or sometimes fresh water were added. After careful shaking the overlaying water was decanted, poured into petri dishes and sorted for living animals under a dissecting microscope. The pipetted animals then were fixed in 4 % formalin seawater solution and kept in small vials.

C. Results

Tetragonicipitidae Lang, 1948
Laophontella Thompson & A. Scott, 1903
Laophontella horrida (Por, 1964) *dentata* nov. subspec.

(Figs. 4–10)

Localities and material. Punta Morales (Locus typicus; 28 August 1990, locality 4); 8 ♀♀, 29 ♂♂, 11 copepodites. Playa Jiménez/Osa (10 September 1990, locality 11); 1 ♂.

8 animals were dissected (4 ♀♀, 4 ♂♂). Holotype female, reg. no. II CR 2 a–1; 6 paratypes (3 ♀♀, 3 ♂♂), reg. no. I CR 31–36. 2. antenna, 1. and 2. maxilla, maxilliped, P 1 and P 3 are drawn from holotype.

Description

Female: Body length of dissected specimens from tip of rostrum to end of furcal rami 0,88–0,92 mm (holotype 0,89 mm). Large parts of the body with a pore-like pattern. Rostrum broad; seen from underside with a suture, from upper side integrated into cephalothorax. Lateral to the scarcely prominent tip there are 2 fine setae (fig. 4 C). Cephalothorax with 2 long conspicuous indentations on distal edge. Following three pereiomeres with lateral indentations (fig. 4 A). Genital double-somite with a lateral sign of subdivision extending to dorsal region. Genital area see fig. 4 B; P 6 with 3 slender setae. Abdominal somites widened laterally. Ventral distal edge of genital double-somite with long hairs. Ventral distal edge of following somite with weak, the one of penultimate somite with distinct toothlike projections. Anal somite dentate on ventral distal edge; one long and one short cone protrude between both furcal rami, respectively. Dorsal distal edges of abdominal somites weakly dentate. Anal operculum ± smooth. Furca complex. Outside protruding a tooth with a seta nearby. Subapically a lobe is developed exhibiting a slender seta distally. Terminal furcal seta about the same length as abdomen; outside 1 short seta is fused with the terminal furcal seta; inside with 1 accompanying slight seta. Dorsal seta standing subapically, bipartite at base. Inner furcal edge dentate. Proximal half of ventral surface with a bulge (figs. 4 A, B; 6 A).

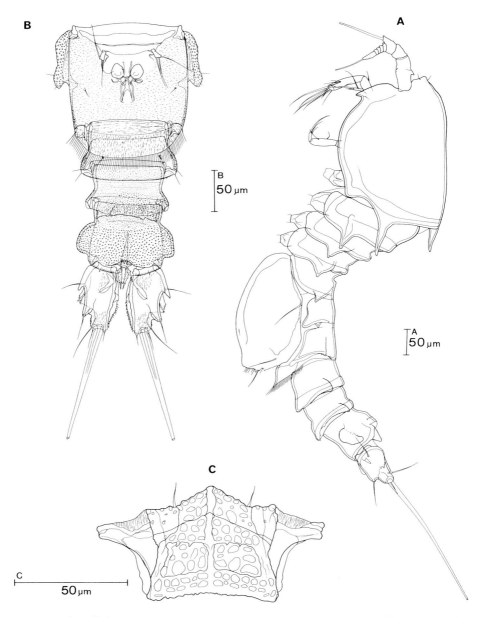

Fig. 4. *Laophontella horrida dentata* nov. subspec. ♀. A. Habitus, lateral side. B. Abdomen, ventral side. C. Rostrum.

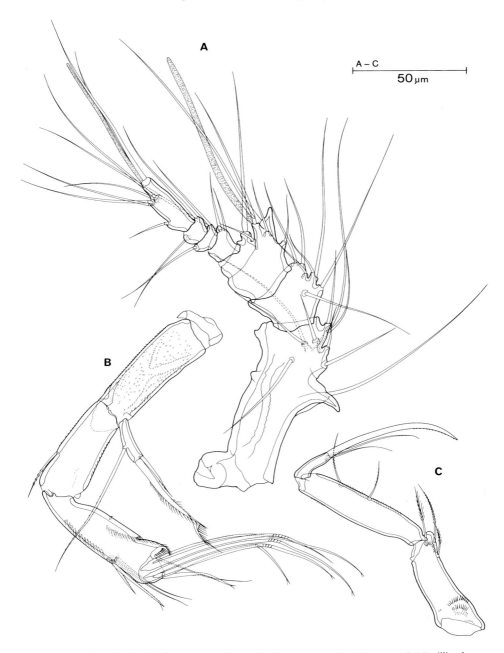

Fig. 5. *Laophontella horrida dentata* nov. subspec. ♀. A. 1. Antenna. B. 2. Antenna. C. Maxilliped.

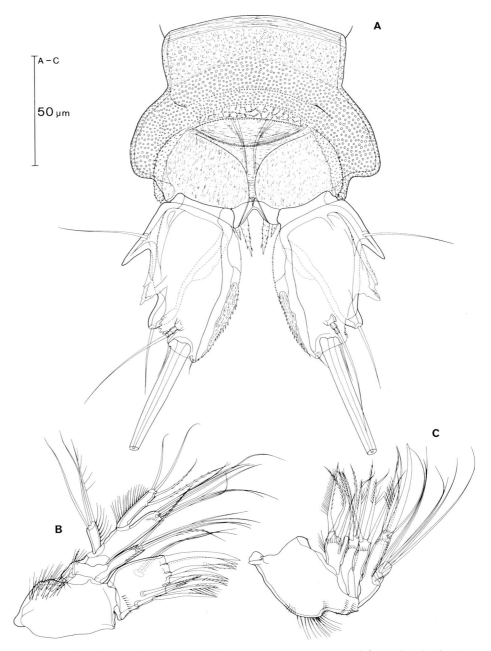

Fig. 6. *Laophontella horrida dentata* nov. subspec. ♀. A. Anal somite and furca, dorsal side. B. 1. Maxilla. C. 2. Maxilla.

1. Antenna (fig. 5 A): 7 (8) segments. 3. segment still with an indicated sign of subdivision. Basal segment elongated, inner edge with a distinct tooth, outer edge with a weak bulge. 3. segment possessing a tooth likewise. Aesthetascs on 3. and last segment.

2. Antenna (fig. 5 B): Basis with small spinules. Exopodite 1-segmented, subapically with 1 long seta, apically with 1 short seta and 1 plumose appendage, the distal part of which is displaced laterally. 1. endopodite-segment with 1 seta, 2. endopodite-segment bearing 4 setae subapically and 7 appendages apically.

Mandible (fig. 7 A): Chewing edge with, in part, strong teeth, subterminally with 2 setalike appendages. Coxa-basis having long hairs and 3 slender setae. Endopodite 1-segmented, elongated, with 2 setae on an incision in the middle and 7 setae apically. Exopodite 2-segmented; basal segment with 2, distal segment with 4 setae.

1. Maxilla (fig. 6 B): Arthrite of precoxa furnished with 7 appendages on distal edge, 2 + 2 setae laterally on inner side and 2 setae on surface. Coxa with 6 setae on endite and 1 seta on exite. Basis, endopodite and exopodite with 8, 4, 3 setae, respectively.

2. Maxilla (fig. 6 C): Syncoxa with three endites. Proximal one deeply incised; both lobes with 1 and 3 setae, respectively. Middle and distal endites bearing 3 setae each, distal pieces of which partly displaced laterally. Basis having 1 claw and 4 setae. Endopodite 2(3?)-segmented with 6 setae altogether and 1 short appendage.

Maxilliped (fig. 5 C): Basis with some rows of spinules and 3 slender setae. 1. endopodite-segment with 2 setae on inner side. 2. endopodite-segment bearing 1 claw and 2 setae.

P 1 (fig. 7 B): Coxa broadened, with several rows of spinules at frontal and caudal surface. Basis with a plumose seta on inner and outer edge. Exopodite 3-segmented. All segments furnished with hairs on edges and spinules on surface. Basal and middle segment each with 1 long seta on outer edge. Distal segment with 4 appendages and 1 rudimentary setule on outer edge. Endopodite 2-segmented. Basal segment prolonged, inner edge with long hairs; outer edge and particularly caudal surface set with spinules. Subapically bearing 1 long plumose seta. Distal segment likewise furnished with hairs and spinules. 2 long appendages apically.

P 2 and P 3 (figs. 8 A, B): Coxa broadened, with several rows of spinules. Basis inside and between both rami with toothlike projections; outer edge with 1 slender seta. Exopodites 3-segmented. Basal segment with 1 long outer appendage; outer edge extended saw-blade-like, with a strong tooth distally. Inner edge with long hairs. Hyaline frill deeply incised. Middle segment likewise with a long outer appendage. Outer edge with rough spinules, inner edge with long hairs

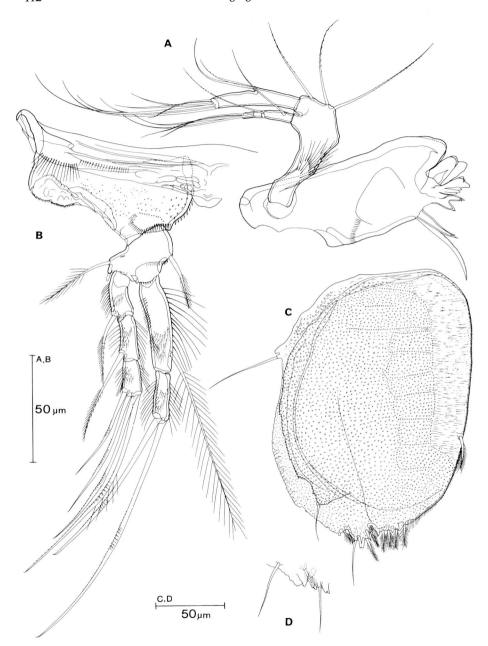

Fig. 7. *Laophontella horrida dentata* nov. subspec. ♀. A. Mandible. B. P 1. C. P 5. D. Distal edge of P. 5, other specimen.

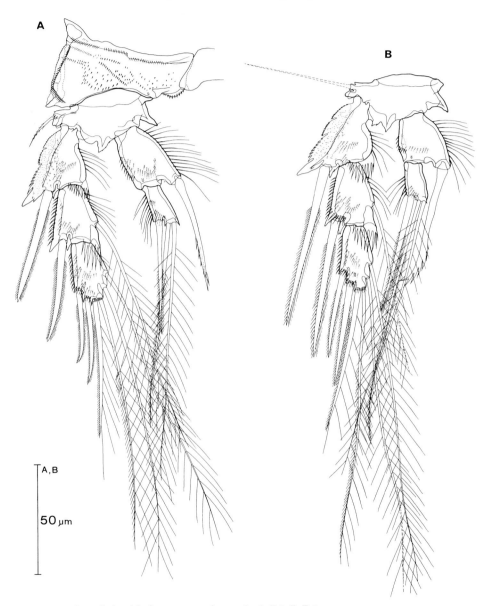

Fig. 8. *Laophontella horrida dentata* nov. subspec. ♀. A. P 2. B. P 3.

and 1 long plumose seta. Hyaline frill deeply incised. Outer edge of distal segment with rough spinules and 2 slender spines. Apically bearing 1 long spinelike appendage and 1 plumose seta. Inner edge with 2 slender plumose setae. Endopodites 2-segmented. Basal segment with long hairs on inner edge and short spinules on outer edge; inner bulge with 1 strong appendage which has denticles terminally. Distal segment with long hairs on outer edge, apically with 3 appendages of different lengths.

P 4 (fig. 9 A): Coxa with several rows of spinules. Basis inside and between both rami with toothlike projections. Outer edge with 1 slender seta. Exopodite 3-segmented. Basal and middle segment saw-blade-like on outer edge; outer appendages shorter than on P 2 and P 3. Inner edge hairy; inner edge of middle segment with an additional short seta. Hyaline frill deeply incised. Distal segment small, furnished with 1 long terminal appendage and 4 short slender setae. Endopodite 2-segmented. Both segments each with 1 plumose appendage.

Seta and spine formula:	Exopodite	Endopodite
P 2	(0.1.222)	(1.021)
P 3	(0.1.222)	(1.021)
P 4	(0.1.122)	(1.010)

P 5 (figs. 7 C, D): Proximal bulge on outer edge bearing 1 seta. Inner edge hairy, with 1 plumose seta. 9–10 setae altogether can be observed subterminally and terminally.

Male: Differing from the female in the following respects:
- Body length 0,77–0,82 mm.
- Ventral distal edges of abdominal somites with indentations. Furcal rami about three times as long as broad. Outer edge with 2 slender setae. Dorsal seta standing subapically; apically with 1 long seta which is accompanied on inner and outer side by one short appendage, respectively (fig. 10 C).
- 1. Antenna (sub-) chirocer.
- Enp. P 2 modified: Distal segment with 1 sickle-shaped seta on inner edge; outer appendage not separated basally (fig. 10 A).
- Enp. P 3 also transformed, particularly the outer appendage on distal segment (fig. 10 B).
- P 4 on basal and middle segment each with 1 strong spine on outer edge terminating bluntly. Distal segment having 1 long, rather strong appendage and 4 slender setae, one of which reaches beyond the length of the strong appendage. Endopodite 2-segmented, each segment with 1 seta (fig. 9 B).

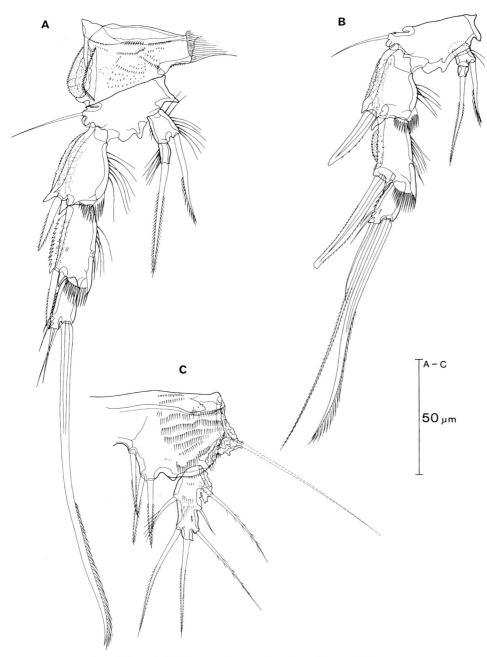

Fig. 9. *Laophontella horrida dentata* nov. subspec. A. P 4 ♀. B. P 4 ♂. C. P 5 ♂.

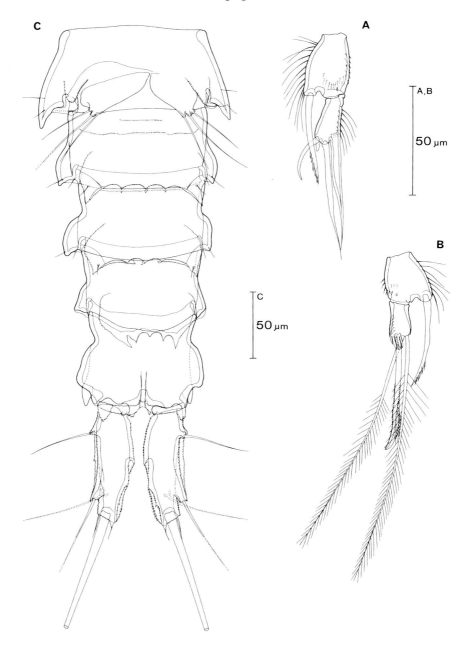

Fig. 10. *Laophontella horrida dentata* nov. subspec. ♂. A. Endopodite P 2. B. Endopodite P 3. C. Abdomen, ventral side.

- P 5: Baseoendopodite with some rows of spinules, 1 seta on outer edge and 3 setae on inner part. Exopodite with 6 setae of different lengths (fig. 9 C).
- P 6 better developed than in female, with 3 setae (fig. 10 C).

Variability. Apart from a somewhat variable manner of body ornamentation and appendages certain setae are occasionally absent (e. g. distal segment exp. P 3 ♀ only with 1 seta on inner edge), or additional setae occur (e. g. benp. P 5 ♂ with 4 setae on inner part).

Etymology. The subspecies name refers to the toothlike projection on outer edge of ♀ furca.

Discussion. Three species of the genus *Laophontella* Thompson & A. Scott, 1903 are known: *L. typica* Thompson & A. Scott, 1903, *L. armata* with both subspecies *L. a. armata* (Willey, 1935) and *L. a. indica* Sewell, 1940, and *L. horrida* (Por, 1964). The ♀ of the latter was first reported from the Israelian coast (POR 1964). At about the same time BODIN (1964) presented a description of ♀ and ♂ of "*Phyllopodopsyllus* sp.?", which had been collected close to Marseille, France. Though both presentations differ to some extent (e. g. 1. antenna, distal segment exp. P 4, intrafurcal spines) in all probability they deal with the same species (BODIN 1967, p. 40).

The specimens of Costa Rica can clearly be identified with *L. horrida*. Nevertheless, some noteworthy deviations may be emphasized, which to my opinion justify the establishment of a new subspecies:
- Body length. The ♀ ♀ of Costa Rica measure about 0,9 mm, the ones of the Mediterranean about 1,2 mm (POR 1964, GUILLE & SOYER 1966).
- Distal segment of enp. P 2 ♂ with a characteristic sickle-shaped seta in the Costa Rican specimens. The Mediterranean ♂ ♂ seemingly have a normally developed seta (see BODIN 1964, Fig. 59).
- Shape of the male furca.
- Female furca in the animals of Costa Rica having a prominent tooth on the outer edge. According to the drawings of POR and BODIN this feature apparently is lacking in the Mediterranean specimens. Therefore the Costa Rican animals are given the rank of the new subspecies *Laophontella horrida dentata*.

L. horrida is known from several localities of the Mediterranean (POR 1964, BODIN 1964, GUILLE & SOYER 1966, BODIOU & SOYER 1973) and from Curaçao (POR 1984). The species seems to prefer coarse sediments (POR 1964: "Gravel?"; BODIN 1964: "psammobiote"; GUILLE & SOYER 1966: "gravelles à Amphioxus; present study). On the other hand POR (1984, see fig. 4 but note the confounded legends!) recorded *Willeyella* = *Laophontella horrida* "from the Curaçao mangal".

Oniscopsis Chappuis, 1954
Oniscopsis robinsoni Chappuis & Delamare Deboutteville, 1956
(Figs. 11–12)

Locality and material. Punta Morales (27 and 28 August 1990, locality 4); 22 ♀♀, 13 ♂♂, several copepodites.
4 ♀♀ and 2 ♂♂ were dissected.
Measured lengths: ♀ 0,52–0,55 mm
♂ 0,41 mm

Discussion. The species was first reported from Bimini (Bahamas) by CHAPPUIS & DELAMARE DEBOUTTEVILLE (1956). Later this species was collected in Mexico (Campeche) and Galapagos (Isola Mosquera) by COTTARELLI et al.

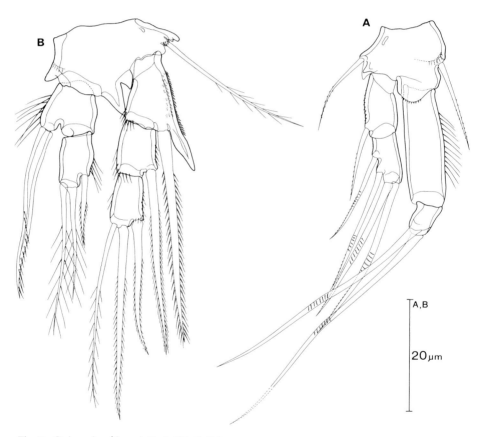

Fig. 11. *Oniscopsis robinsoni.* ♀. A. P 1. B. P 2.

(1985) and obtained from various islands of the Galapagos Archipelago by MIELKE (1989).

The animals of Punta Morales largely correspond with the available descriptions. It was already pointed out that ornamentation and appendages of specimens from different Galapagos beaches as well as individuals from one and the same beach may show slight variations. The animals from the only locality known so far from Costa Rica also show intraspecific deviations: (1) Inside and outside of their bases the endopodites of P 2 and P 3 are bordered by a toothlike projection; the inner tooth may be reduced occasionally. (2) Enp. P 4 ♀ reduced to a varying extent. (3) The toothlike projection on basal segment of exp. P 4 may be developed rather more weakly than is shown in fig. 12 B. (4) Furca and anal operculum are also shaped somewhat variably.

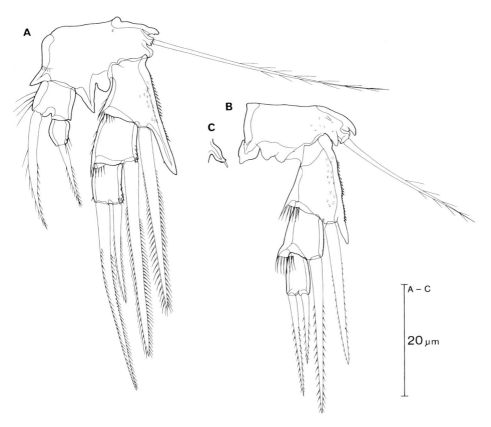

Fig. 12. *Oniscopsis robinsoni.* ♀. A. P 3. B. P 4. C. Rudimentary enp. P 4, other side.

Phyllopodopsyllus T. Scott, 1906
Phyllopodopsyllus setouchiensis Kitazima, 1981

(Figs. 13–17)

Localities and material. Punta Morales (27 and 28 August 1990, locality 4); 259 ♀♀, 256 ♂♂, a number of copepodites, several precopulatory couples. Playa Manzanillo (7 September 1990; locality 3); several adult specimens. Playa Jiménez/Osa (10 September 1990, locality 11); 1 ♀. Playa Blanca/Osa (10 September 1990, locality 12); 7 ♀♀, 2 ♂♂.
  25 animals were dissected (15 ♀♀, 10 ♂♂).
  Measured lengths: ♀ 0,46–0,79 mm
           ♂ 0,46–0,56 mm

Discussion. Within the taxon *Phyllopodopsyllus* the species *P. briani* Petkovski, 1955; *P. setouchiensis* Kitazima, 1981; *P. mielkei mielkei* Kunz, 1984 (the state of *P. mielkei californicus* Kunz, 1984 is unclear, because of the differing setation of the distal segment on exopodite P 4 ♀); *P. petkovskii* Kunz, 1984; *P. galapagoensis* Mielke, 1989 with all probability constitute a monophyletic partial group. Due to their evident variability, it is extremely difficult to decide whether they are all true species. The delimitation of the species given by the authors are only based on gradual deviations and differences in proportions. The species cited above may represent (1) local races of one species, in which some differences of features have been manifested according to the great spatial distance of the localities; (2) subspecies of *P. briani*; (3) reproductively isolated species with great conformity in their external appearance. A decision only based on morphological evidence is not possible.

The treated animals (figs. 13–17) from various localities of Costa Rica can easily be identified as *P. setouchiensis*, which KITAZIMA (1981) has recorded from "the sandy beach of Mukaishima Island in the Inland Sea of Japan". The most conspicuous difference concerns the bulbous basic part of the terminal furcal seta. However, this part is likewise not uniform in the partial populations of Costa Rica (figs. 16 C, D, F). Otherwise there are only some minor divergences. There are also striking similarities with *P. mielkei mielkei*. Probably the populations of Japan (KITAZIMA 1981), Hawaii (KUNZ 1984) and Costa Rica (present study) belong to one and the same species.

The other three representatives of *Phyllopodopsyllus* mentioned above reveal some features, which prove – at least provisionally – their existence as species: (1) *P. briani*: weak development of the projection on second segment of 1. antenna; exopodite and endopodite P 1 more or less of the same length; "normal" structure of the terminal furcal seta. (2) *P. petkovskii*: weak development of the projection on second segment of 1. antenna; "normal" structure of the terminal furcal seta. Both species are separated only by gradual differences. (3) *P. galapagoensis:* 1. antenna with 8 segments; terminal furcal seta stunted.

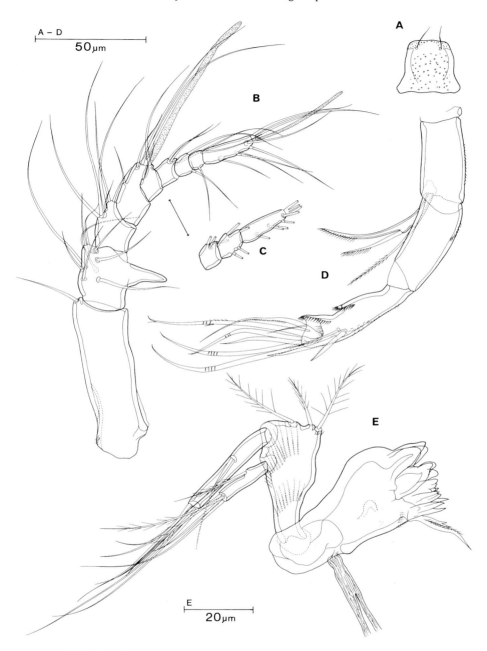

Fig. 13. *Phyllopodopsyllus setouchiensis*. ♀. A. Rostrum. B. 1. Antenna. C. Distal part of 1. Antenna, other specimen. D. 2. Antenna. E. Mandible.

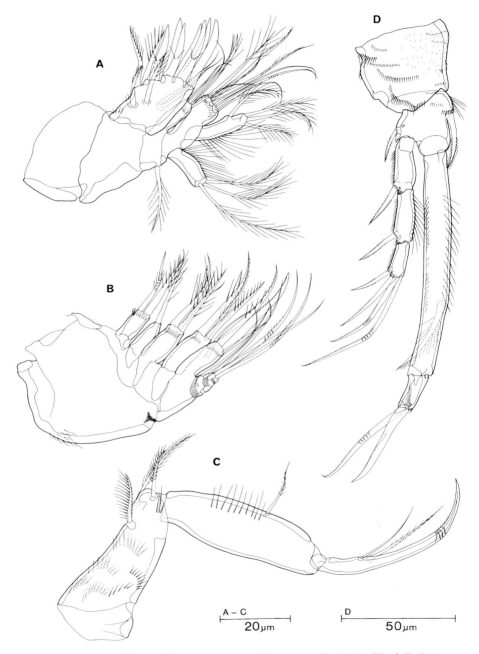

Fig. 14. *Phyllopodopsyllus setouchiensis.* ♀. A. 1. Maxilla. B. 2. Maxilla. C. Maxilliped. D. P. 1.

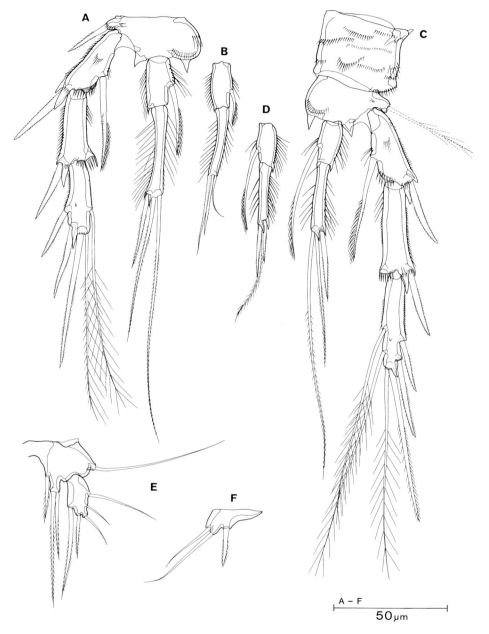

Fig. 15. *Phyllopodopsyllus setouchiensis*. A. P 2 ♀. B. Endopodite P 2 ♂. C. P 3 ♀. D. Endopodite P 3 ♂. E. P 5 ♂. F. P 6 ♂.

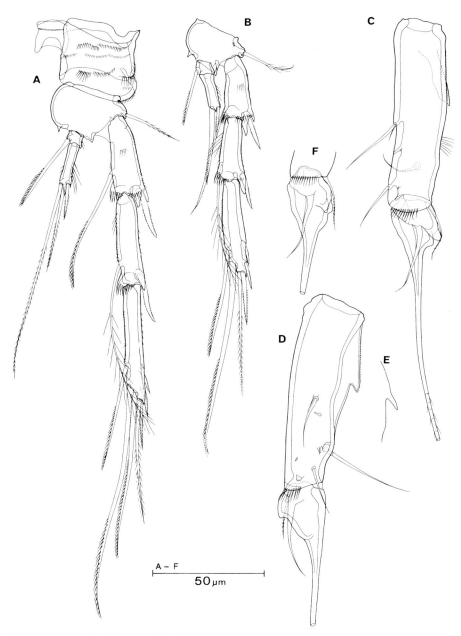

Fig. 16. *Phyllopodopsyllus setouchiensis*. A. P 4 ♀. B. P 4 ♂. C. Furca, ventral side ♀. D. Furca, lateral side ♀. E. Dorsal edge of furca ♀. F. Basal part of terminal furcal seta ♀ (C–F from different specimens).

Within the partial populations of Costa Rica some variable features can be ascertained concerning, above all, the furca ♀ (figs. 16 C–F) and 1. antenna ♀ (figs. 13 B, C). It was already pointed out (see discussion about *P. galapagoensis* in MIELKE 1989) that the different segment number of the 1. antenna must be interpreted as being of subordinate significance. The "7. segment" of the 9-segmented state is still subdivided, whereas in the 8-segmented state the dividing line is reduced. These facts can clearly be demonstrated with respect to the animals of Costa Rica. The specimens of Punta Morales and the Peninsula Osa

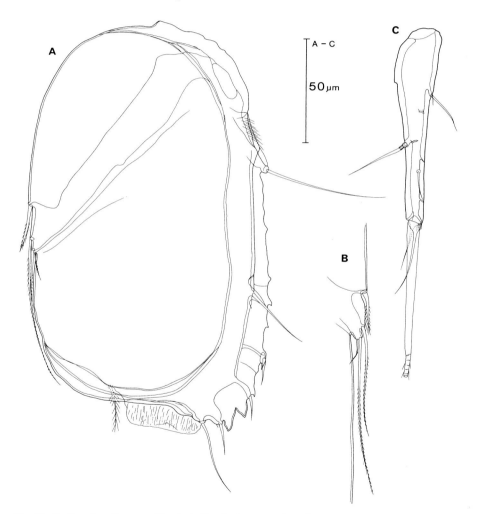

Fig. 17. *Phyllopodopsyllus setouchiensis*. A. P 5 ♀. B. Inner edge of P 5 ♀. C. Furca, lateral side ♂.

have 9 segments, the ones of Manzanillo 9 or 8. Here an intermediate state can be observed.

The apparent clear delimitation between *P. galapagoensis* and the animals of Costa Rica (and the other mentioned species) becomes problematic when those individuals are included, which were discussed in MIELKE (1989, "Bemerkung", p. 161). Here, too, the terminal furcal seta is well developed (specimens of Fernandina I, 2 and Isabela II, 5). The only remaining differences are the segment number of 1. antenna and the structure of the basic part of the terminal furcal seta. On the other hand, the animals of San Cristóbal (XIII, 2) clearly differ from the ones of Costa Rica. They rather resemble *P. briani*: Terminal furcal seta "normal"; toothlike projection on second segment of 1. antenna only weakly developed; exopodite and endopodite of P 1 more or less of the same length. Therefore I hesitated to class these animals with *P. galapagoensis*. It cannot be clarified at present whether or not the Galapagos populations represent several species or only one species, from which, after its immigration into the archipelago, morphologically deviating partial populations have formed by speciation. Furthermore, the question concerning the place of origin of the species, which immigrated into the Galapagos Archipelago, is still unsettled. The coast of Costa Rica or Central America, respectively, can be taken into consideration.

P. setouchiensis partly occurs in high abundance in the different localities. Especially in the beach of Punta Morales it represents one of the most frequent copepod species. Its presence on the Pacific coast (Punta Morales and Península Osa) and on the Atlantic coast (Manzanillo) proves *P. setouchiensis* to be an amphi-American species.

Phyllopodopsyllus ancylus nov. spec.
(Figs. 18–21)

Locality and material. Punta Morales (Locus typicus; 28 August 1990, locality 4); 3 ♀♀, 2 ♂♂.

All 5 animals were dissected. Holotype female, reg. no. I CR 39; 4 paratypes (2 ♀♀, 2 ♂♂), reg. no. I CR 38, 40, 41, 42. 2. antenna, mandible, 1. and 2. maxilla, maxilliped, P 1, P 4 are drawn from holotype.

Description

Female: Body length from tip of rostrum to end of furcal rami 0,44–0,49 mm (holotype 0,47 mm). Rostrum more or less square, bearing two fine setae subapically (fig. 18 A). Genital double-somite with a subdivision line which is obviously furnished with fine hairs. Other distal edges of abdominal somites seemingly bare. Furca irregularly shaped. Outer part with 2 slender

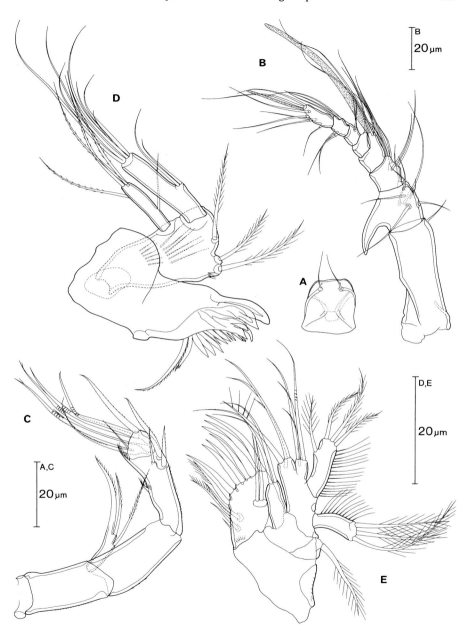

Fig. 18. *Phyllopodopsyllus ancylus* nov. spec. ♀. A. Rostrum. B. 1. Antenna. C. 2. Antenna. D. Mandible. E. 1. Maxilla.

setae. At about three quarters of the length of dorsal edge a seta inserts which is bipartite at base. Terminal furcal seta accompanied by 2 slender setae the outer one is fused with the terminal seta (fig. 21 A).

1. Antenna (fig. 18 B): 8 segments. 2. segment with a strong, caudally bent tooth. 4. and 8. segment each with an aesthetasc.

2. Antenna (fig. 18 C): Basis with small setules on anterior edge. Exopodite 1-segmented, with 3 appendages, outer one basally fused with the exopodite. 1. endopodite-segment only with a few setules on anterior edge. 2. endopodite-segment with distinct spinules on anterior and posterior edges; subapically with 3, apically with 7 appendages of different lengths.

Mandible (fig. 18 D): Chewing edge of precoxa with a row of teeth of different thickness and 2 setalike appendages. Coxa-basis with 3 plumose setae and long spinules on the surface. Endopodite 1-segmented, with 2 + 7 setae. Exopodite carrying 5 setae altogether.

1. Maxilla (fig. 18 E): Arthrite of precoxa with 9 appendages on distal edge and 2 setae each on inner and outer side. Coxa with 4 (5?) setae on endite and 1 seta on exite. Basis, endopodite and exopodite with 8, 4, 3 setae, respectively.

2. Maxilla (fig. 19 A): Syncoxa with 4 endites, carrying 2, 1, 3, 3 setae. Basis having 4 appendages. Endopodite 2- or 3-segmented with 6 setae altogether.

Maxilliped (fig. 19 B): Basis with 3 plumose setae and rows of spinules on the surface. 1. endopodite-segment with 1 seta and long spinules on inner edge. 2. endopodite-segment having 1 claw and 2 slender setae.

P 1 (fig. 19 C): Coxa with rows of spinules on frontal and caudal (not drawn) surface. Basis with 1 small seta on outer edge and 1 strong plumose appendage on inner part. Exopodite 3-segmented. All segments furnished with spinules on outer edge, middle segment with long hairs on inner edge. Distal segment with 4 slender appendages. Endopodite 2-segmented. Proximal segment with long spinules on inner edge and short spinules on outer edge; subapically, at about two thirds of the length 1 plumose seta inserts. Distal segment with spinules and 2 claw-like appendages.

P 2 and P 3 (figs. 20 A, B): Coxa with some rows of spinules on frontal and caudal (not drawn) surface. Basis inside and between both rami with toothlike projections. Outer edge bearing 1 seta which is clearly longer in P 3 than in P 2. Exopodites 3-segmented. All segments with spinules on outer edge and toothlike projections on distal outer edge. Basal segment with a row of spinules on caudal surface and 1 strong seta on inner edge, middle segment only with spinules on inner edge. Both segments with a deeply incised hyaline frill on distal inner edge. Distal segment with 4 appendages. Endopodites 2-segmented. Basal segment with 1 seta on inner edge. Distal segment apically with 2 toothlike projections and 3 appendages.

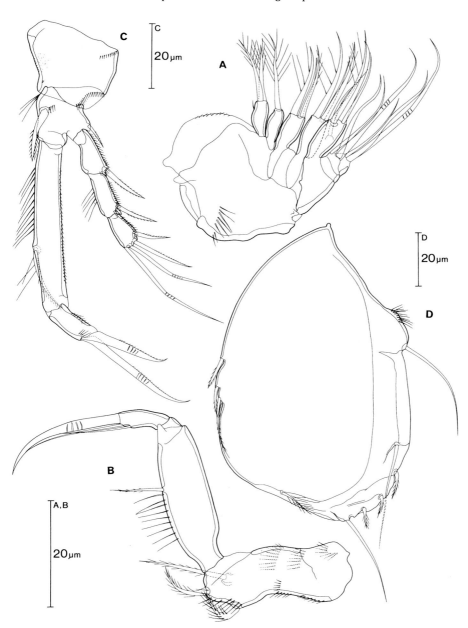

Fig. 19. *Phyllopodopsyllus ancylus* nov. spec. ♀. A. 2. Maxilla. B. Maxilliped. C. P 1. D. P 5.

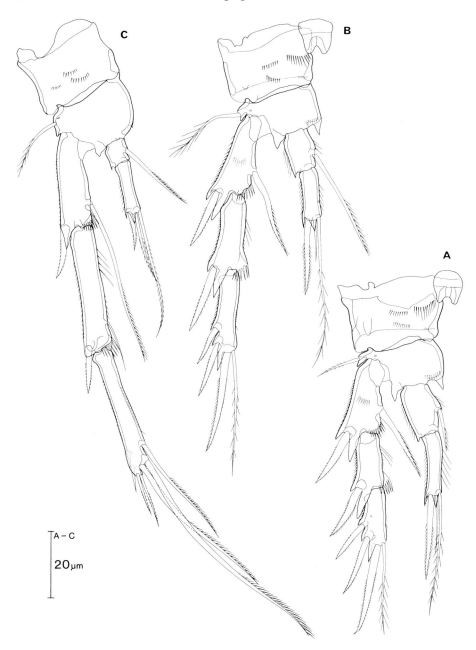

Fig. 20. *Phyllopodopsyllus ancylus* nov. spec. ♀. A. P 2. B. P 3. C. P 4.

P 4 (fig. 20 C): Coxa with some rows of spinules. Basis with 1 slender seta on outer edge; distal edge toothlike prolonged between both rami; subdistally, on inner edge with a weak denticle. Exopodite 3-segmented. Basal segment with an outer spine and 1 long, slender inner seta. Hyaline frill on distal inner edge deeply incised. Middle segment as well with 1 outer spine and a deeply incised hyaline frill; inner seta dwarfed. Distal segment with 7 appendages altogether. Endopodite 2-segmented. Proximal segment short, with 1 inner seta. Distal segment with 3 appendages of different lengths.

Seta and spine formula:		Exopodite	Endopodite
	P 2	(1.0.022)	(1.021)
	P 3	(1.0.022)	(1.021)
	P 4	(1.1.322)	(1.021)

P 5 (fig. 19 D): Proximal bulge on outer edge with 1 slender seta. Subapically and apically with 2 long slender setae and 4 short plumose setae. Inner edge armed with 4 setae of different lengths.

Male: Differing from the female in the following respects:
- Body length 0,43 and 0,45 mm.
- Furcal rami about five times longer than broad (measured at the broadest part proximally). Outer edge with 2 setae and 1 accompanying setule. Dorsal seta situated at 3/5 of the length. Distal furcal seta accompanied by a slender outer seta which is basally fused with the main seta and a short inner seta (fig. 21 G).
- 1. Antenna (sub-) chirocer.
- Distal segment of enp. P 2 with 3 setae, inner one is slightly bent inwards; outer seta not separated basally (fig. 21 B).
- Distal segment of enp. P 3 with a strong tooth on distal outer edge (fig. 21 C).
- Distal outer edge of middle segment P 4 blade-like in contrast to female P 4. Distal segment exp. P 4 only with 6 setae. Distal segment enp. P 4 only with 2 appendages, inner one strong and bent inwards (fig. 21 D).
- Benp. P 5 with 1 outer seta and 3 plumose setae on inner part. Exopodite with 5 setae (fig. 21 E).
- P 6 with 2 appendages of different length (fig. 21 F).

Variability. One P 4 of one male exhibits an interesting inversion of setal armature. The slender outer seta of basis is obviously developed on proximal segment of exopodite, the outer spine of which apparently occurs at the basis.

Etymology. The specific name is derived from the Greek *ancylos* (=

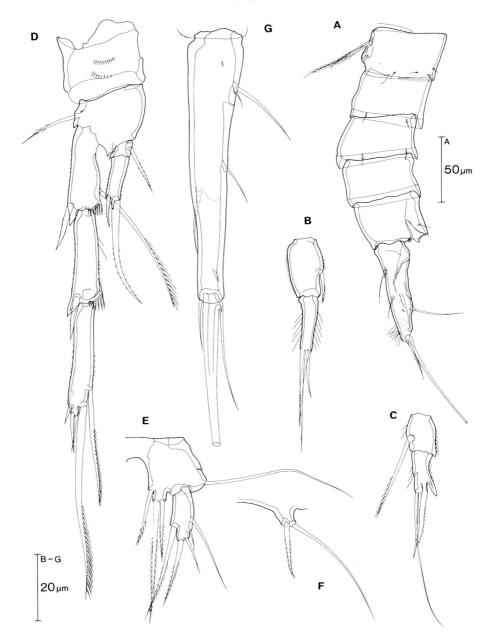

Fig. 21. *Phyllopodopsyllus ancylus* nov. spec. A. Abdomen, lateral side ♀. B. Endopodite P 2 ♂. C. Endopodite P 3 ♂. D. P 4 ♂. E. P 5 ♂. F. P 6 ♂. G. Furca, ventral side ♂.

crooked, curved) and refers to the conspicuous, caudally bent tooth an 2. segment of the 1. antenna.

Discussion. *P. ancylus* shows similarities with several *Phyllopodopsyllus* species, especially with *P. pauli* Crisafi, 1959 and *P. danielae* Bodin, 1964. Nevertheless, the new species can be clearly distinguished from all other known representatives of the taxon by the seta and spine formula P 2 – P 4 and the structure of 1. antenna and furca.

Phyllopodopsyllus carinatus nov. spec.
(Figs. 22–27)

Locality and material. Playa Manzanillo (Locus typicus; 7 September 1990, locality 3); frequent species.

12 animals were dissected. Holotype female, reg. no. I CR 51; 6 paratypes (3 ♀ ♀, 3 ♂ ♂), reg. no. I CR 45, 46, 49, 52, 53, 54. Rostrum and 1. antenna, 2. antenna, 2. maxilla, maxilliped, P 3 – P 5 are drawn from holotype.

Description

Female: Body length from tip of rostrum to end of furcal rami 0,54–0,67 mm (holotype 0,61 mm). Rostrum about as long as broad, subapically with two fine setae (fig. 23 A). Last pereion somite and abdominal somites seemingly with delicate denticles on dorsal distal edge. Genital double-somite and following somite with fine spinules ventrolaterally. Penultimate and ultimate somites having fine spinules on the whole ventral distal edge. Many of the spinules bear a hairlike structure (not drawn) obviously not belonging to the hyaline frill but rather representing an algal thread. Genital double-somite with a dividing mark. Operculum with spinules. Furca irregularly shaped and almost twice as long as broad. Outer part with 2 slender setae. Distal outer edge prolonged and carrying spinules; inner part of furca with long spinules. Dorsal seta bipartite at base inserting distally. A medio-ventral keel-like structure is obliquely directed to inner side. Distal furcal edge furnished with 3 appendages: outer one plumose and broadened basally; inner seta slender. Middle terminal furcal seta with a thick proximal part having a cone-like structure (figs. 22 A, B; 24 D).

1. Antenna (fig. 23 A): 9 segments. 2. segment with a projection on outer edge. 4. and last segment each with an aesthetasc.

2. Antenna (fig. 23 B): Basis and both endopodite-segments with spinules on anterior edge. Exopodite 1-segmented, with 3 appendages, outer one basally fused with the exopodite. 2. endopodite-segment subapically with 4, apically with 7 appendages.

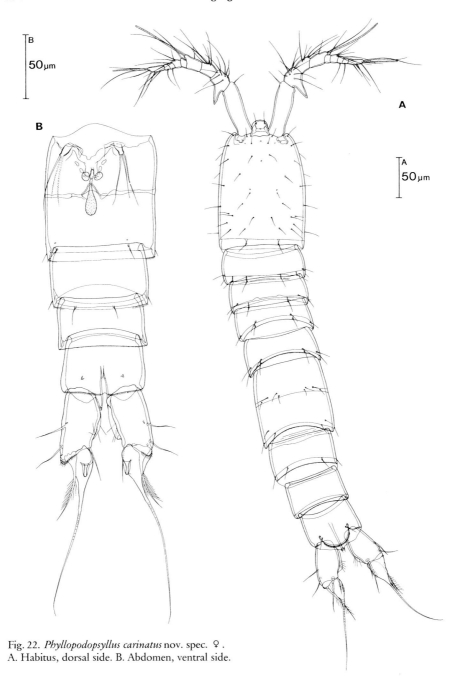

Fig. 22. *Phyllopodopsyllus carinatus* nov. spec. ♀.
A. Habitus, dorsal side. B. Abdomen, ventral side.

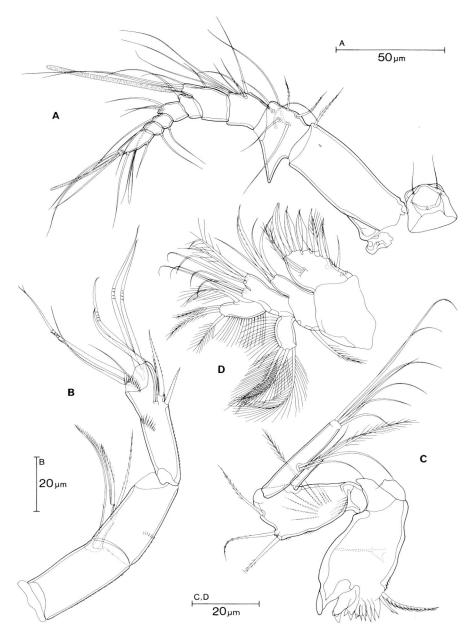

Fig. 23. *Phyllopodopsyllus carinatus* nov. spec. ♀. A. Rostrum and 1. antenna. B. 2. Antenna. C. Mandible. D. 1. Maxilla.

Mandible (fig. 23 C): Chewing edge of precoxa with several teeth and two setalike appendages. Coxa-basis furnished with long spinules and 3 plumose setae. Endopodite 1-segmented, with 2 setae at a break on half of the length and 7 slender setae apically. Exopodite with 5 setae altogether.

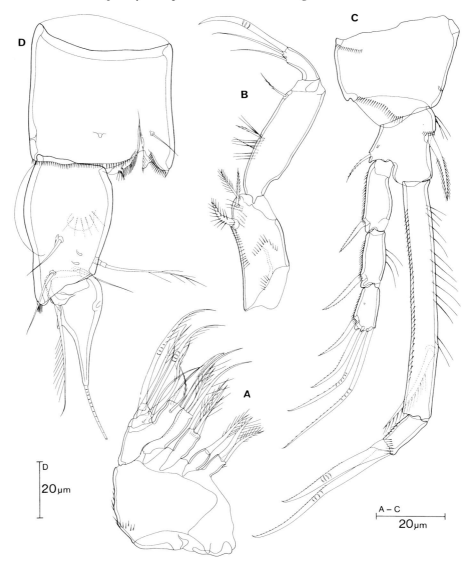

Fig. 24. *Phyllopodopsyllus carinatus* nov. spec. ♀. A. 2. Maxilla. B. Maxilliped. C. P. 1. D. Anal somite and furca, lateral side.

1. Maxilla (fig. 23 D): Arthrite of precoxa with 9 appendages on distal edge, 2 setae on inner part and 2 slender setae on surface. Coxa furnished with 5 setae on endite and 1 seta on exite. Basis with a notch and 8 setae altogether. Endopodite and exopodite with 4 and 3 plumose setae, respectively.

2. Maxilla (fig. 24 A): Syncoxa with 4 endites which have 2, 1, 3, 3 setae. Basis with 4 appendages. Endopodite 2- or 3-segmented, with 6 setae and apparently 1 small hook basally.

Maxilliped (fig. 24 B): Basis with rows of spinules and 3 plumose appendages. 1. endopodite-segment with 2 setae and long spinules on inner edge. 2. endopodite-segment carrying 1 claw and 2 slender setae.

P 1 (fig. 24 C): Coxa with rows of spinules. Basis with 1 short outer seta and 1 strong plumose seta on inner side which bears some additional hairlike spinules. Exopodite 3-segmented. All segments with spinules on outer edge; middle segment with long hairs on inner edge. Distal segment with 4 slender appendages. Endopodite 2-segmented. Proximal segment with long hairs on inner edge and short spinules on outer edge; subapically on inner side bearing 1 plumose seta. Distal segment short with spinules and 2 claw-like appendages apically.

P 2 and P 3 (figs. 25 A, B): Coxa with some rows of spinules. Basis inside and between both rami with toothlike projections; outside a slender seta inserts. Exopodites 3-segmented. All segments with spinules on outer edges which taper off to pointed projections. Proximal segment with 1 inner plumose seta. Distal inner edge of proximal and middle segments with a deeply incised hyaline frill. Distal segment with 4 appendages. Endopodites 2-segmented. Basal segment with 1 strong inner seta, distal segment bearing 3 appendages of different lengths; outer short appendage basally framed by toothlike projections.

P 4 (fig. 26 A): Coxa with some rows of spinules. Basis inside and between both rami with toothlike projections; outer edge having 1 slender seta. Exopodite 3-segmented. Basal segment with a long inner seta which is pectinate on distal half. Middle segment without inner seta. Both segments with a deeply incised hyaline frill on distal inner edge. Distal segment furnished with 7 appendages. Endopodite 2-segmented. Proximal segment with 1 inner seta. Distal segment with 3 appendages apically.

Seta and spine formula:	Exopodite	Endopodite
P 2	(1.0.022)	(1.021)
P 3	(1.0.022)	(1.021)
P 4	(1.0.322)	(1.021)

P 5 (fig. 26 B): Proximal outer edge hairy. Nearby on proximal outer edge a bulge protrudes carrying 1 slender seta. Distal outer part having 1 long seta and 2 spinelike appendages. Distal edge with 2 setae. Inner edge subdistally with 1 plumose seta; proximal and middle part of inner edge furnished with 5 setae, one being very long.

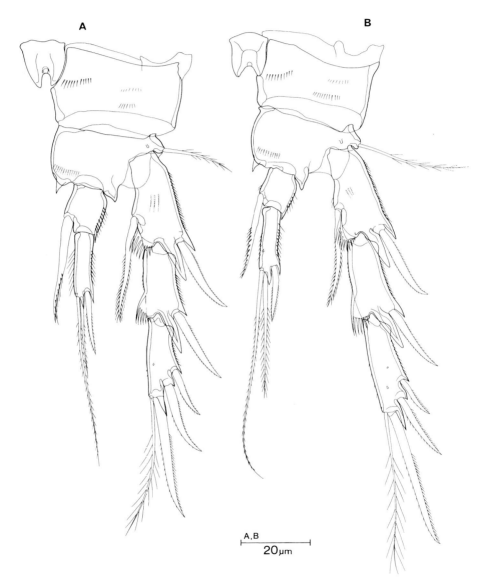

Fig. 25. *Phyllopodopsyllus carinatus* nov. spec. ♀. A. P 2. B. P 3.

Six representatives of the Tetragonicipitidae 139

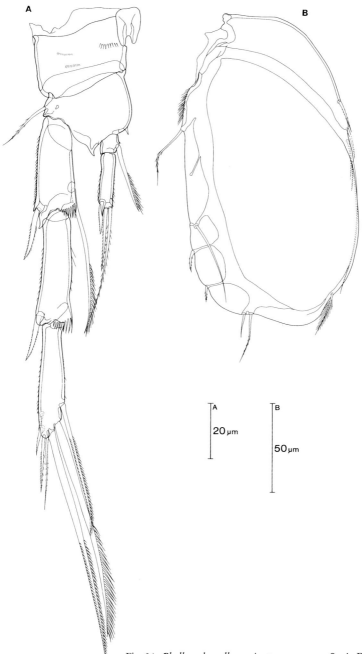

Fig. 26. *Phyllopodopsyllus carinatus* nov. spec. ♀. A. P 4. B. P 5.

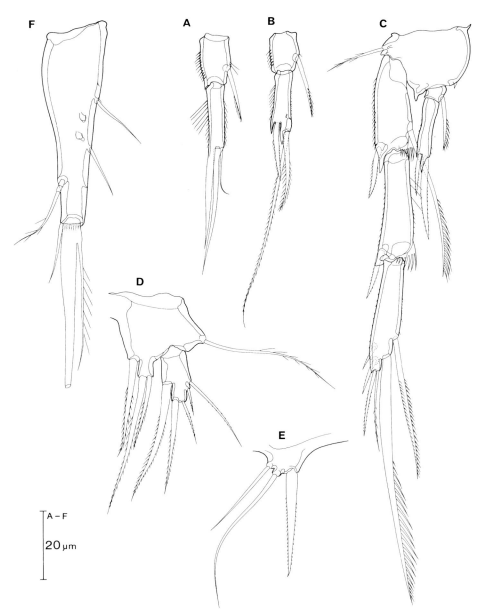

Fig. 27. *Phyllopodopsyllus carinatus* nov. spec. ♂. A. Endopodite P 2. B. Endopodite P 3. C. P 4. D. P 5. E. P 6. F. Furca, dorsal side.

Male: Differing from the female in the following respects:
- Body length 0,41–0,49 mm.
- Distal ventral edge of 2. and 3. abdominal somites with a continuous row of spinules.
- Furcal rami about three times longer than broad (measured at the broadest part proximally). Outer edge having 2 setae. Dorsal seta inserting subdistally. Distally with 3 setae: inner one short, outer one hairy and basally fused with the terminal main seta (fig. 27 F).
- 1. Antenna subchirocer.
- Distal segment of enp. P 2 with 3 appendages, inner one of which is slightly bent inwards; outer appendage basally fused with the segment (fig. 27 A).
- Enp. P 3 weakly modified (fig. 27 B).
- Distal outer edge of middle segment exp. P 4 more distinct than in female. Distal segment with 6 appendages. Distal segment of enp. P 4 only with 2 setae, inner one strong (fig. 27 C).
- Benp. P 5 with 1 outer slender seta and 3 setae on inner part. Exopodite with 5 setae, innermost one longest (fig. 27 D).
- P 6 with an inner spinelike appendage and 2 slender setae (fig. 27 E).

Etymology. The specific name is derived from the Latin *carina* (= keel) and refers to the ventral keel-like structure on the furca of the female.

Discussion. The new species exhibits the same seta and spine formula P 2 – P 4 as several other *Phyllopodopsyllus* species (*P. berrieri* Monard, 1936; *P. danielae* Bodin, 1964; *P. hibernicus* (Roe, 1955); *P. laspalmensis* Marinov, 1973; *P. mossmani* T. Scott, 1912). Distinctive features are the 1. antenna and the furca as well as two additional characteristics that to my knowledge have not been known from any other *Phyllopodopsyllus* species: The presence of 2 setae on 1. endopodite segment of Maxilliped and of 5 setae on inner edge of P 5 ♀.

Phyllopodopsyllus gertrudi Kunz, 1984
costaricensis nov. subspec.

(Figs. 28, 29)

Localities and material. Playa Manzanillo (Locus typicus; 7 September 1990, locality 3); several specimens. Playa Blanca/Osa (10 September 1990, locality 12); 1 ♀.
6 ♀ ♀ and 5 ♂ ♂ were dissected. Holotype female, reg. no. I CR 56; 5 paratypes (2 ♀ ♀, 3 ♂ ♂), reg. no. I CR 55, 57–60.
Measured lengths: ♀ 0,54–0,60 mm
♂ 0,41–0,45 mm
Body length of the single ♀ found in Playa Blanca/Osa only 0,42 mm.

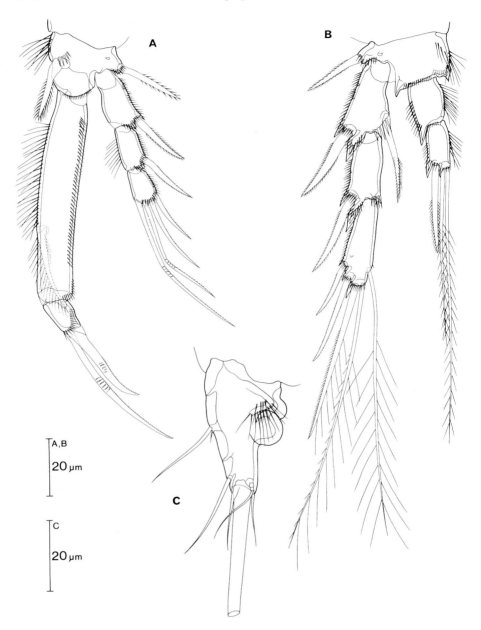

Fig. 28. *Phyllopodopsyllus gertrudi costaricensis* nov. subspec. ♀. A. P 1. B. P 2. C. Furca, dorsal side.

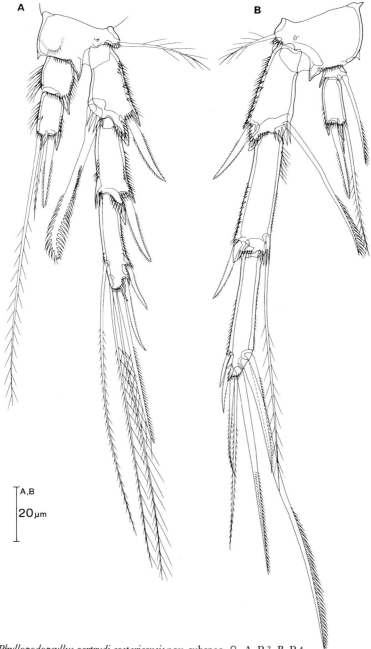

Fig. 29. *Phyllopodopsyllus gertrudi costaricensis* nov. subspec. ♀. A. P 3. B. P 4.

Discussion. The material collected in Costa Rica contained only one *Phyllopodopsyllus* species lacking a lateral projection on 2. segment of 1. antenna. Without doubt this species is related to *P. aegypticus* Nicholls, 1944, *P. angolensis* Kunz, 1984 and *P. gertrudi* Kunz, 1984. Apart from the 1. antenna the three species share some characteristic features, e. g. structure of the mouth parts, the typical strong inner appendage on basal segment exp. P 3 and the furcal set-up. Though the Costa Rican animals exhibit the identical seta and spine formula P 2 – P 4 as *P. angolensis*, there are at least two important features that support a closer relationship with *P. gertrudi*: (1) length ratio exp.: enp. of P 1; (2) structure of the female furca. The present specimens differ from the ones which KUNZ (1984) reported from some Hawaiian Islands especially in (1) the setation of distal segment exp. P 3 (present animals 222, Hawaiian specimens 122); (2) the setation of distal segment exp. P 4 ♂ (5 and 6 setae, respectively); (3) length ratio of both apical setae on distal segment enp. P 2 and P 3 (different and equal lengths, respectively).

Because of these differences a new subspecies is erected for the Costa Rican animals: *P. gertrudi costaricensis*. It represents another example for so-called amphi-American species demonstrating the close relationship of many taxa occurring on both sides of the Central American isthmus.

Drawings are presented from P 1 – P 4 and furca ♀ (figs. 28, 29).

Zusammenfassung

Von diversen Stränden Costa Ricas werden sechs Vertreter der Tetragonicipitidae vorgestellt.

Mit *Laophontella horrida dentata* wird eine neue Unterart errichtet.

Die Individuen zweier weiterer Arten werden mit *Oniscopsis robinsoni* Chappuis & Delamare Deboutteville, 1956 und *Phyllopodopsyllus setouchiensis* Kitazima, 1981 identifiziert.

Phyllopodopsyllus ancylus und *P. carinatus* werden als neue Arten beschrieben.

Die neue Unterart *Phyllopodopsyllus gertrudi costaricensis* wird mit der Nominat-Unterart *P. gertrudi gertrudi* Kunz, 1984 verglichen.

Acknowledgements

Above all I want to thank Dr. Manuel M. Murillo, director of the "Centro de Investigación en Ciencias del Mar y Limnología" (CIMAR), University of Costa Rica, San José, for his kind invitation and the possibility to use the facilities of his institute.

Dr. José A. Vargas was my adviser and supported me in many ways during all the time.

Prof. Jenaro Acuña and Prof. Ricardo Soto took me to their study sites at the Caribbean coast and the Península Osa, respectively. Thus I could collect material from some hardly accessible localities.

Furthermore I want to thank all the staff of the CIMAR in San José and the marine laboratory in Punta Morales.

The research trip was made possible by a financial support of the Deutsche Forschungsgemeinschaft (Mi 218/2–3).

References

BODIN, P. (1964): Recherches sur la systématique et la distribution des Copépodes Harpacticoides des substrats meubles des environs de Marseille. Rec. Trav. St. Mar. End. **33**, 107–183.
– (1967): Catalogue des nouveaux Copépodes Harpacticoides marins. Mém. Mus. Nat. Hist. nat. **50**, 1–75.
BODIOU, J.-Y. & J. SOYER (1973): Sur les Harpacticoides (Crustacea, Copepoda) des sables grossiers et fins graviers de la région de Banyuls-sur-Mer. Rapp. Comm. int. Mer Médit. **21**, 657–659.
BRUSCA, R. C. & E. W. IVERSON (1985): A Guide to the Marine Isopod Crustacea of Pacific Costa Rica. Rev. Biol. Trop. **33** (Supl. 1), 1–77.
CHAPPUIS, P. A. & C. DELAMARE DEBOUTTEVILLE (1956): Etudes sur la faune interstitielle des îles Bahamas récoltée par Madame Renaud-Debyser. I. Copépodes et Isopodes. Vie Milieu **7**, 373–396.
COLLADO, C., D. DEFAYE, B. H. DUSSART & C. H. FERNANDO (1984 a): The freshwater Copepoda (Crustacea) of Costa Rica with notes on some species. Hydrobiologia **119**, 89–99.
COLLADO, C., C. H. FERNANDO & D. SEPHTON (1984 b): The freshwater zooplankton of Central America and the Caribbean. Hydrobiologia **113**, 105–119.
COTTARELLI, V., P. E. SAPORITO & A. C. PUCCETTI (1985): Il genere *Oniscopsis* Chappuis (Crustacea, Copepoda, Harpacticoida): Osservazioni morfologiche e faunistiche. Boll. Mus. civ. St. nat. Verona **12**, 257–272.
DUSSART, B. & C. H. FERNANDO (1985 a): Tropical freshwater Copepoda from Papua, New Guinea, Burma, and Costa Rica, including a new species of *Mesocyclops* from Burma. Can. J. Zool. **63**, 202–206.
DUSSART, B. & C. H. FERNANDO (1985 b): Remarks on two species of copepods in Costa Rica, including a description of a new species of *Tropocyclops*. Crustaceana **50**, 39–44.
GUILLE, A. & J. SOYER (1966): Copépodes Harpacticoides de Banyuls-sur-Mer. 4. Quelques formes des gravelles á Amphioxus. Vie Milieu **17**, 345–387.
KITAZIMA, Y. (1981): Three new species of the genus *Phyllopodopsyllus* (Copepoda, Harpacticoida) from the Inland Sea of Japan. Publ. Seto Mar. Biol. Lab. **26**, 393–424.
KUNZ, H. (1984): Beschreibung von sechs *Phyllopodopsyllus*-Arten (Copepoda, Harpacticoida) vom Pazific. Mitt. Zool. Mus. Univ. Kiel **2**, 11–32.
LAGUNA, J. E. (1990): Shore barnacles (Cirripedia, Thoracica) and a revision of their provincialism and transition zones in the Tropical Eastern Pacific. Bull. Mar. Sci. **46**, 406–424.
LÖFFLER, H. (1972): Contribution to the limnology of High Mountain Lakes in Central America. Int. Revue ges. Hydrobiol. **57**, 397–408.
MAURER, D., J. VARGAS & H. DEAN (1988): Polychaetous Annelids from the Gulf of Nicoya, Costa Rica. Int. Revue ges. Hydrobiol. **73**, 43–59.
MIELKE, W. (1989): Interstitielle Fauna von Galapagos. XXXVI. Tetragonicipitidae (Harpacticoida). Microfauna Marina **5**, 95–172.
POR, F. D. (1964): A study of the Levantine and Pontic Harpacticoida (Crustacea, Copepoda). Zool. Verh. Rijksmus. Natuurl. Hist. Leiden **64**, 1–128.
– (1984): Notes on the benthic Copepoda of the mangal ecosystem. In: Por, F. D. & J. Dor (eds.): Hydrobiology of the Mangal. Dr. W. Junk Publishers, The Hague, p. 67–70.
REID, J. W. (1990): *Canthocamptus (Elaphoidella) striblingi*, new species (Copepoda: Harpacticoida) from Costa Rica. Proc. Biol. Soc. Wash. **103**, 336–340.

VARGAS, J. A. (1987): The benthic community of an intertidal mud flat in the Gulf of Nicoya, Costa Rica. Description of the community. Rev. Biol. Trop. **35**, 299–316.
– (1988): A survey of the meiofauna of an Eastern Tropical Pacific intertidal mud flat. Rev. Biol. Trop. **36**, 541–544.
VOORHIS, A. D., C. E. EPIFANIO, D. MAURER, A. I. DITTEL & J. A. VARGAS (1983): The estuarine character of the Gulf of Nicoya, an embayment on the Pacific coast of Central America. Hydrobiologia **99**, 225–237.

Dr. Wolfgang Mielke
II. Zoologisches Institut und Museum der Universität Göttingen
Berliner Straße 28, D-3400 Göttingen

Neue interstitielle Polychaeten (Hesionidae, Dorvilleidae) aus dem Litoral des Golfs von Bengalen

Wilfried Westheide

Inhaltsverzeichnis

Abstract 147
A. Einleitung 148
B. Artbeschreibungen 148
    Hesionides bengalensis nov. sp. 148
    Microdorvillea phuketensis nov. sp. 150
    Parapodrilus indicus nov. sp. 154
Zusammenfassung 156
Danksagung 156
Literatur 157

New interstitial Polychaeta (Hesionidae, Dorvilleidae) from the littoral of the Bay of Bengal

Abstract

Three species of interstitial meiofauna polychaetes new to science are described from intertidal and shallow subtidal sandy sediments. *Hesionides bengalensis* n. sp. (Hesionidae) was found in sandy beaches of the Island of Phuket (Thailand); some specimens from South Andaman formerly considered to belong to *Hesionides indooceanica* Westheide & Rao were included into the new species. The dorvilleid *Parapodrilus indicus* n. sp. was discovered in a beach at Mandapam, Gulf of Mannar; it shows close relationships to the European *Parapodrilus psammophilus* Westheide, which was the only species of this genus hitherto known. *Microdorvillea phuketensis* n. sp. from sandy patches between coral reefs on Phuket and South Andaman is another dorvilleid of a hitherto monotypic genus.

A. Einleitung

Aus Sandstränden des Golfs von Bengalen sind bereits zahlreiche Meiofauna-Polychaeten bekannt (ALIKUNHI 1951; BANSE 1959; RAO & GANAPATI 1967, 1968; RAO 1975; JOUIN & RAO 1987; WESTHEIDE 1990 a, 1990 b). Die für diesen Lebensraum besonders charakteristische Gattung *Hesionides* Friedrich (Hesionidae) hat hier offensichtlich ein Zentrum der Evolution und ist mit mehr Arten vertreten als in jeder anderen Region der Erde (WESTHEIDE & RAO 1977; RAO 1978). Die vorliegende Arbeit stellt eine weitere Art dieser Gattung vor, dazu zwei neue Arten bisher monotypischer Gattungen der Familie Dorvilleidae. Die Tiere sind Teil weltweiter Aufsammlungen von Meiofauna-Polychaeten und wurden bei kurzen Aufenthalten in Thailand und Indien gefunden. Methodik, siehe WESTHEIDE (1990 a,b).

B. Artbeschreibungen

Hesionides bengalensis nov. sp.

(Abb. 1)

Hesionides indooceanica Westheide & Rao, 1977, partim.

Fundorte: – (1) Insel Phuket (Thailand); Andamanen See; Karon Beach (98° 18' 0, 07° 50' N); exponierter Brandungsstrand, mittlerer Hang: Locus typicus. Phuket Marine Center (98° 24' 0, 07° 47' N), kleiner wenig exponierter Strand aus Korallensand, oberer Hang (26. 10. 1987). (2) Andaman Inseln, South Andaman (Indien), Andamanen See (siehe WESTHEIDE & RAO 1977, p. 282–284).

Material: – 5 fixierte Exemplare von Phuket, 3 von den Andamanen. Holotypus ist ein Exemplar mit 22 Borstensegmenten von Phuket (Reference Collection, Phuket Marine Center Nr. PMBC 7269).Paratypen im Zoologischen Museum der Universität Hamburg (Nr. P 20471) und in der Sammlung des Autors.

Färbung. – Durchsichtig, bräunlich.

Körpergröße. – Länge bis 1,9 mm; Breite vorn ca. 100 µm, hinten ca. 70 µm (ohne Parapodien). 19–26 Borstensegmente.

Vorderende. – Mit der für die Gattung charakteristischen Zahl von 11 fadenförmigen Anhängen mit den typischen knotenförmigen Anschwellungen und der breiten Basis (Abb. 1 A,B). Dorsale und ventrale prostomiale Tentakel von ungefähr gleicher Länge (ca. 75 µm); unpaarer dorsaler Anhang deutlich länger. Die drei Paar Tentakelcirren nehmen von vorn nach hinten an Länge zu; 3. Tentakelcirren bis zu 180 µm lang.

Borstensegmente. – Fast gleichartige Segmente, die jedoch nach hinten an Breite deutlich abnehmen (Abb. 1 A); in den letzten Segmenten nur unvollständige Parapodien. Notopodien mit Dorsalcirren, die die Neuropodien und ihre Borsten weit überragen. Bündel aus feiner gerader Acicula und zwei einfachen, leicht gebogenen Borsten unterschiedlicher Länge; die längere mit 5 (auch 6) (Abb. 1 D) sägeförmigen Zähnchen, die kürzere mit wahrscheinlich nur 3 Zähnchen (Abb. 1 E), die schwer zu erkennen sind. Neuropodium mit fingerförmigem Cirrus; meist 5 zusammengesetzte Borsten mit Endglied von sehr unterschiedlicher Länge, zweispitzig (Abb. 1F–H); eine feine, gebogene (Abb. 1 K) und eine größere gerade, zugespitzte Acicula (Abb. 1 J).

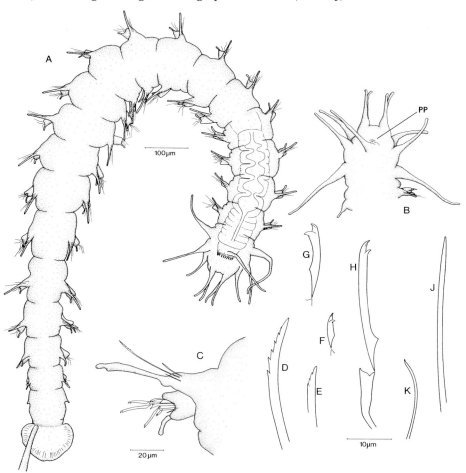

Abb. 1. *Hesionides bengalensis* nov. sp. A. Habitus. B. Vorderende und Penispapillen (pp). C. Mittleres Parapodium, von hinten gesehen. D–K. Borsten aus dem 12. Borstensegment. D, E. Notopodiale Borsten. F–H. Zusammengesetzte neuropodiale Borsten. J, K. Neuropodiale Aciculae. 10 µm-Maßstab für alle Borstenzeichnungen. Alle Zeichnungen mit Zeichenapparat.

Hinterende. – Pygidium mit einheitlichem, fächerförmigen Anallappen, der breiter als das letzte Segment ist (Abb. 1 A). Zwei fadenförmige Analcirren, die bei fixierten Tieren meist fehlen.

Innere Organisation. – Langer muskulöser Pharynx, der bis in das 5. Borstensegment reicht; Öffnung mit spitz zulaufenden Papillen umgeben; durch 6 tiefe Einschnürungen in Kammern unterteilt (Abb. 1 A).

Geschlechtsorgane. – Männchen mit kurzen paarigen Penispapillen, die direkt vor der Basis des medianen Tentakels auf dem Prostomium liegen (Abb. 1 B).

Diskussion. – In den Sandstränden von Phuket und den Andamanen kommen mehrere *Hesionides*-Arten nebeneinander vor, darunter an beiden Fundorten die weltweit verbreiteten *H. arenaria* Friedrich und *H. gohari* Hartmann-Schröder. Die neue Art ist durch einen einheitlichen, nicht in der Mitte geteilten Schwanzlappen charakterisiert, den innerhalb der Gattung nur noch *H. unilamellata* Westheide und *H. indooceania* Westheide & Rao besitzen. Von den beiden Arten unterscheidet sich die neue Art deutlich durch die Form ihrer notopodialen Borsten, von *H. indooceanica* auch noch durch die geringe Größe der notopodialen Aciculae. Besonders charakteristisch für *H. bengalensis* nov. spec. ist die starke Faltung des Pharynx, die in dieser Weise bei keiner anderen *Hesionides*-Art vorkommt. Auch Form und Lage der Penispapillen sind abweichend, allerdings kann dieses Merkmal bisher nur an fünf Arten verglichen werden (WESTHEIDE 1984, Fig. 8; WESTHEIDE 1987).

In die Beschreibung von *Hesionides indooceanica* wurden auch drei Individuen aufgenommen, die nicht die kräftigen notopodialen Aciculae aufwiesen (WESTHEIDE & RAO 1977) und zunächst als Varianten dieser Art geführt wurden. Die in der Beschreibung geäußerte Vermutung, daß es sich um Vertreter einer getrennten Art handeln könnte, läßt sich jetzt bestätigen. Nachuntersuchungen haben gezeigt, daß diese Individuen zu der hier beschriebenen Art gehören. Ihre in WESTHEIDE & RAO (1977) abgebildeten notopodialen Borsten (Abb. 4 P,Q) besitzen einige Zähne weniger als gezeichnet und stimmen so auch in diesen Merkmalen mit den Phuket-Tieren überein.

Microdorvillea phuketensis nov. sp.

(Abb. 2)

Fundorte: – (1) Insel Phuket (Thailand); Andamanen See. Westküste, Patong Beach (09° 17' 0, 07° 59' N). Detritusreicher Sand vor einem Korallenriff bei Niedrigwasser: Locus typicus (24. 10. 1987). (2) Andaman Inseln, South Andaman (Indien), Andamanen See; Südküste, in der Nähe von Chiriatapu; feiner sauberer Sand im flachen Wasser bei Niedrigwasser vor einem kleinen Korallenriff (22. 1. 1988).

Material: – Zwei lebende Individuen und drei fixierte von Phuket; 11 fixierte Individuen von den Andamanen. Holotypus ist ein Exemplar mit 16 Borstensegmenten (Reference Collection, Phuket Marine Center Nr. PMBC 7270). Paratypen im Zoologischen Museum der Universität Hamburg (Nr. P 20470) und in der Sammlung des Autors.

Färbung. – Weißlich, durchsichtig.

Körpergröße. – Länge bis zu 1,5 mm; Breite ca. 65 µm, bzw. 140 µm (mit Parapodien). Maximal 21 Borstensegmente.

Vorderende. – Prostomium ei- bis kugelförmig, vorn breit gerundet, fast ebenso lang wie breit (Abb. 2 A). Mit zwei Paar Anhängen: (1) latero-ventral befestigte und schräg nach hinten gerichtete Palpen (Länge ca. 60 µm), deutlich gegliedert in einen zylindrischen Palpophor und einen wenig kürzeren, aber deutlich dünneren Palpostyl; (2) dorsal, ungefähr auf gleicher Höhe kegelförmige ungegliederte Antennen (Länge ca. 45 µm). Steife sensorische Cilien auf Antennen, Palpen und Apex des Prostomiums. Zwei Cilienringe auf dem Prostomium, davon einer vor den Antennen. Keine Augen. Nuchalorgane wahrscheinlich seitlich in der Furche zwischen Prostomium und Rumpf. Zwei parapodienlose Körperabschnitte, breiter als lang; der erste mit zwei, der hintere mit einem vollständigen Cilienring.

Borstensegmente. – Fast gleichartige Borstensegmente, breiter als lang; die vorderen auffällig kürzer als die mittleren und hinteren. Parapodien uniram, deutlich vom Körperstamm abgesetzt. Ohne Dorsalcirrus, aber mit kleinem lappenförmigen Ventralcirrus. Wahrscheinlich auf allen Segmenten ein Cilienring. Die Tiere können schnell und wendig nur mit ihren Cilien schwimmen; die Parapodien liegen dabei dem Körper eng an.

Borstenausstattung der Parapodien: 2–3 einfache und 3–4 zusammengesetzte Borsten. Einfache Borsten knieförmig mit gezähntem subdistalen Abschnitt, unterschiedlich in Größe und Form: in den vorderen Segmenten kräftiger, deutlich gebogen, z. T. sichelförmig, mit kürzerem glatten distalen Teil und starken Zähnchen (Abb. 2 F–J); in den mittleren und hinteren Parapodien gestreckter und feiner (Abb. 2 K). Zusammengesetzte Borsten mit einfachem einspitzigen und wahrscheinlich glatten Endteil von unterschiedlicher Länge (Abb. 2 L–P).Klingen und Schäfte feiner, sowie Endteile länger in den mittleren und hinteren Segmenten (Abb. 2 O). Schäfte distal zweispitzig (Abb. 2 N); auch dreispitzige Schäfte wurden beobachtet (Abb. 2 P). Eine gerade, gleichmäßig sich verjüngende Acicula (Abb. 2 E).

Hinterende. – Hinterste Segmente kleiner, das letzte ohne Parapodien. Pygidium zweimal so lang wie breit; mit unpaarem medianen papillenartigen Anhang und zwei fingerförmigen terminalen Analcirren (Länge bis ca. 100 µm) (Abb. 2 A).

Kieferapparat. – Flügelförmige, fast glattrandige Mandibeln, ohne bräunliche

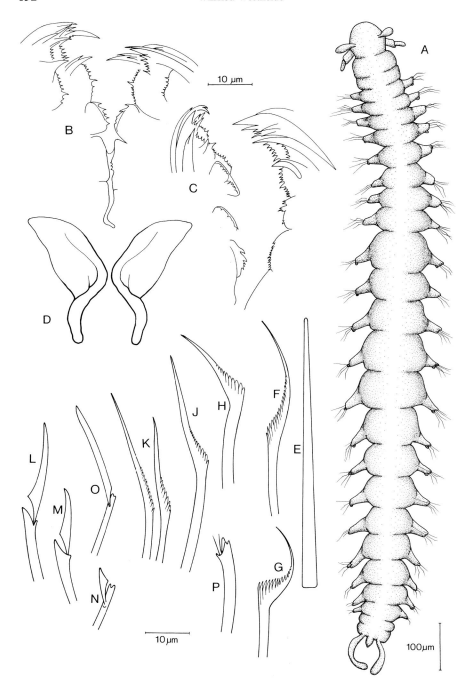

Sklerotisierung (Abb. 2 D). Schwer erkennbare Maxillenelemente (Abb. 2 B, C); in 4 Reihen angeordnet: schwach differenzierte posteriore Grundplatte, davor innere Reihen aus je 4 Platten und je eine außen liegende Platte; Plattenränder mit kleinen spitzen Zähnchen; anterior krallenförmige Elemente, deren Zahl nicht genau erkannt werden konnte.

Geschlechtsorgane. – Borstensegmente 12, 14, 15, 17 und 18 eines Tieres mit je einer dotterreichen Oocyte.

Diskussion. – Innerhalb der immer größer werdenden Zahl kleiner Dorvilleiden gehört die vorliegende Art zu einer Gruppe von Arten, die 2 Paar kurze prostomiale Anhänge besitzen, von denen die Palpen zweigliedrig sind. Diese Merkmale findet man bei *Meiodorvillea chilensis* (Hartmann-Schröder, 1965), *Ophryotrocha longidentata* Josefson, 1975, *O. lobifera* Oug, 1978 und den Gattungen *Pettiboneia* Orensanz, *Microdorvillea* Westheide & von Nordheim und *Westheideia* Wolf.

Schon die Kiefer zeigen, daß die neue Art nicht zur Gattung *Ophryotrocha* gehört. *Pettiboneia* hat stärker differenzierte Parapodien mit Dorsalcirren, andere Borstentypen und eine größere Zahl von Maxillenelementen. Ähnliche Unterschiede bestehen auch zu *Westheideia*, deren einzige Art subbirame Parapodien, Gabelborsten und andere Kieferelemente besitzt (WOLF 1986). *Meiodorvillea chilensis*, deren systematische Stellung überprüft werden sollte, hat 4 Analcirren und Gabelborsten (HARTMANN-SCHRÖDER 1965). Der Vergleich mit dieser Art wird im übrigen durch fehlende Kenntnis der Kiefer behindert. Die größten Übereinstimmungen bestehen zu der *Microdorvillea otagensis* aus Neuseeland (WESTHEIDE & VON NORDHEIM 1985). Das Vorderende ist identisch, ebenso das Pygidium bis auf den deutlicher abstehenden medianen Anhang. Die Parapodien besitzen einen kleinen Ventralcirrus, der bei *M. otagensis* nicht gefunden wurde. Die Borstenbewaffnung besteht in beiden Arten aus je einem Typ einfacher und zusammengesetzter Borsten. Die einfachen Kapillarborsten von *M. otagensis* unterscheiden sich nicht grundsätzlich von den knieförmigen Borsten in den mittleren und hinteren Segmenten der vorliegenden Art. Die Kieferapparate zeigen einen besonders hohen Grad an Übereinstimmung, vor allem in der Struktur und Anordnung der Maxillenelemente. Die neue Art wird daher in die Gattung *Microdorvillea* gestellt. Dies bedeutet allerdings, daß die Gattungs-

◀ Abb. 2. *Microdorvillea phuketensis* nov. sp. A. Habitus. B,C. Maxillen-Elemente des Kieferapparates von zwei verschiedenen Tieren. D. Mandibeln. E. Acicula, 2. Borstensegment. F–J. Gekniete einfache Borsten aus vorderen Segmenten. K. Gekniete einfache Borsten aus einem mittleren Segment. L–N. Zusammengesetzte Borsten aus 3. und 4. Borstensegment. O. Zusammengesetzte Borste aus einem hinteren Borstensegment. P. Dreispitziger Schaft einer zusammengesetzten Borste. 10 µm-Maßstab gilt für alle Borstenzeichnungen. Alle Zeichnungen mit Zeichenapparat.

diagnose leicht verändert werden muß. Statt „unirame Parapodien ohne dorsale und ventrale Cirren" muß es nun heißen „unirame Parapodien ohne Dorsalcirren".

Parapodrilus indicus nov. sp.
(Abb. 3)

Fundort: – Tamilnadu (Indien), Golf von Mannar; Mandapam, Sandstrand nördlich des Flüchtlingslagers (ca. 9° 16' N, 79° 08' O) (weitere Einzelheiten des Fundortes, siehe BANSE 1959): Locus typicus (3. 2. 1988).

Material: – 11 fixierte Exemplare. Holotypus ist das Dauerpräparat eines nicht ganz vollständigen Tieres (Zoologisches Museum der Universität Hamburg Nr. P 20469).

Färbung. – Durchsichtig, farblos – weißlich.

Körpergröße. – Länge 470–630 µm; Breite ca. 60 µm (ohne Parapodien).

Vorderende. – Längliches, vorn stark gerundetes Prostomium, mit seitlichen Einschnürungen; mehrere strahlenförmig nach vorn gerichtete Bündel steifer Cilien; seitlich 2 Wimperringe zu erkennen, von denen der vordere wahrscheinlich dorsal vollständig ist. Von den beiden folgenden tonnenförmigen Rumpfabschnitten ist der vordere deutlich länger und durch eine Einschnürung leicht gegliedert; jeder Abschnitt mit Wimperring.

Borstensegmente. – Die für die Gattung charakteristischen 4 Borstensegmente dreiteilig, nach hinten an Länge zunehmend. Stummelförmige, nach schräg hinten gerichteten Parapodien mit je einer geraden, weit in den Körper reichenden und vorn spitz zulaufenden Acicula und drei einfachen, glatten, leicht geschwungenen, ebenfalls spitz zulaufenden Borsten (Abb. 3 B). Je ein Wimperring vor den Parapodien. Zahlreiche feine sensorische Cilien auf den Parapodien.

Hinterende. – Auf das letzte Borstensegment folgt ein relativ langes borstenloses, durch Einschnürungen gegliedertes Segment mit in der Mitte liegendem Wimperring. Fast ebenso langes Pygidium, mit drei Analcirren: zwei laterale, ventrad gerichtete Anhänge (Länge 22–29 µm) und ein wesentlich kürzerer ventraler medianer Anhang (Abb. 3 C). Pygidium mit Wimperring in der vorderen Hälfte und zahlreichen sensorischen Cilien am Hinterrand und auf den Anhängen.

Geschlechtsorgane. – Mindestens ein Tier mit einer relativ großen Eizelle (Länge ca. 75 µm), die vom 4. Borstensegment in das nachfolgende Segment hineinragt (Abb. 3 A).

Diskussion. – Die durch nur 4 Borstensegmente charakterisierte Gattung *Parapodrilus* Westheide, 1965 gehört zu jener Gruppe kleiner Dorvilleiden mit

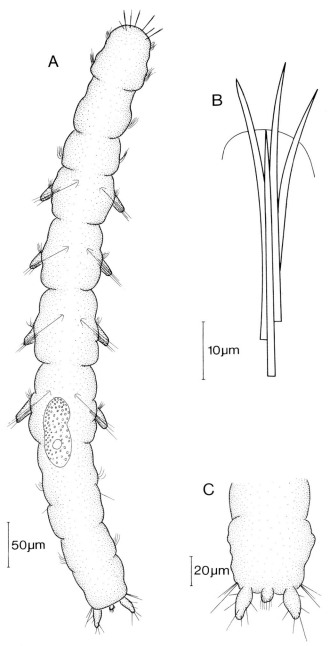

Abb. 3. *Parapodrilus indicus* nov. sp. A. Habitus. B. Borstenbündel eines Parapodiums. C. Pygidium mit Anhängen. Zeichnungen mit Zeichenapparat.

stark neotener („progenetic") Organisation, von denen in den letzten Jahren zahlreiche neue Arten aus dem Sandlückensystem beschrieben wurden (z. B. WESTHEIDE 1984; WESTHEIDE & VON NORDHEIM 1985; WOLF 1986). *Parapodrilus psammophilus* wurde in einem Sandstrand der Nordseeinsel Sylt gefunden (WESTHEIDE 1965) und auch an der französischen Atlantikküste nachgewiesen (WESTHEIDE 1972/73). Nur RISER (1984) meldet die Gattung noch von Neuseeland und entdeckte sie auch an der US-Atlantikküste (persönliche Mitteilung). Das Auffinden einer weiteren Art von der Küste Südindiens belegt erneut die weltweite Verbreitung interstitieller Polychaetengattungen aus Gezeitensandstränden.

Die hier beschriebenen Tiere wurden beim Aufsammeln lebend gesehen, konnten aber nur im fixierten Zustand beschrieben werden. Epidermisdrüsen und genaue Einzelheiten der Bewimperung lassen sich deshalb nicht zum Vergleich mit der europäischen Art heranziehen. Die völlig glatten Borsten unterscheiden die neue Art jedoch hinreichend deutlich von der europäischen *P. psammophilus* mit ihren fein gesägten Borsten. Auch die Analcirren differieren: *P. indicus* hat lange laterale und einen kurzen mittleren Anhang; bei *P. psammophilus* ist dagegen der mittlere Anhang der längste. Wahrscheinlich bestehen auch Unterschiede in der Länge des hinteren borstenlosen Segments und des Pygidiums.

Zusammenfassung

Bei Aufsammlungen aus dem Litoral verschiedener Regionen des Golfs von Bengalen wurden drei neue interstitielle Polychaeten gefunden. *Hesionides bengalensis* nov. sp. (Hesionidae) lebt in exponierten Sandstränden der Inseln Phuket (Thailand) und South Andaman (Indien). *Parapodrilus indicus* nov. sp. und *Microdorvillea phuketensis* nov. sp. sind Dorvilleidae aus bisher monotypischen Gattungen, die in Sandstränden Südindiens, bzw. zwischen Korallenstöcken auf Phuket und South Andaman vorkommen.

Danksagung

Die Aufsammlungen wurden mit Hilfe und durch Gastfreundschaft verschiedener Personen und Institutionen ermöglicht: Direktor und Mitarbeiter des Phuket Marine Center, insbesondere Herr Anuwat Nateewathana auf Phuket, Thailand; Dr. G. C. Rao, Port Blair, South Andaman; Prof. Dr. T. J. Pandian und Mitarbeiter seines Labors, Madurai Kamaraj University, Madurai, Indien;

das Bundesministerium für Forschung und Technologie, Bonn und Frau E. Hongsernant, Köln. Ihnen allen sei hier herzlich gedankt.

Literatur

ALIKUNHI, K. H,. 1951. On the reproductive organs of *Pisione remota* (Southern), together with a review of the family Pisionidae (Polychaeta). Proc. Ind. Acad. Sci., Sec. B., **33**, 14–31.

BANSE, K., 1959. On marine Polychaeta from Mandapam (South India). J. Mar. biol. Ass. India **1**, 165–177.

HARTMANN-SCHRÖDER, G., 1965. Zur Kenntnis des Sublitorals der chilenischen Küste unter besonderer Berücksichtigung der Polychaeten und Ostracoden. Teil II. Die Polychaeten des Sublitorals. Mitt. Hamb. Zool. Mus. Inst., Ergänzungsband zu **62**, 59–305.

JOUIN, C. & RAO, G. C., 1987. Morphological studies on some Polygordiidae and Saccocirridae (Polychaeta) from the Indian Ocean. Cah. Biol. mar. **28**, 389–402.

RAO, G. C., 1975. The interstitial fauna in the intertidal sands of Andaman and Nicobar group of islands. J. mar. biol. Ass. India **17**, 116–128.

RAO, G. C., 1988. On a new species of *Hesionides* (Polychaeta: Hesionidae) from Orissa coast, India. Bull. zool. Surv. India **1**, 271–274.

RAO, G. C. & GANAPATI, P. N., 1967. On some interstitial polychaetes from the beach sands of Waltair coast. Proc. Ind. Acad. Sci. **65**, sec. B., 10–15.

RAO, G. C. & GANAPATI, P. N., 1968. On some archiannelids from the beach sands of Waltair coast. Proc. Ind. Acad. Sci **67**, sec. B, 24–30.

RISER, N. W., 1984. General observations on the intertidal interstitial fauna of New Zealand. Tane **30**, 239–250.

WESTHEIDE, W., 1965. *Parapodrilus psammophilus* nov. gen. nov. spec., eine neue Polychaeten-Gattung aus dem Mesopsammal der Nordsee. Helgol. wiss. Meeresunters. **12**, 207–213.

WESTHEIDE, W., 1972–73. Nouvelles récoltes d'annélides interstitielles dans les plages sableuses du Bassin d'Arcachon. Vie Milieu **23**, 365–370.

WESTHEIDE, W., 1984. The concept of reproduction in polychaetes with small body size: Adaptation in interstitial species. Fortschr. Zool. **29**, 265–287.

WESTHEIDE, W., 1987. The interstitial polychaete *Hesionides pettiboneae* n. sp. (Hesionidae) from the U.S. east coast and its transatlantic relationships. Bull. Biol. Soc. Wash. **7**, 131–139.

WESTHEIDE, W., 1990 a. A new genus and species of the Syllidae (Annelida, Polychaeta) from South India. Zool. Scr. **19**, 165–167.

WESTHEIDE, W., 1990 b. A hermaphroditic *Sphaerosyllis* (Polychaeta: Syllidae) with epitokous genital chaetae from intertidal sands of the Island of Phuket (Thailand). Can. J. Zool. **68**, 2360–2363.

WESTHEIDE, W. & VON NORDHEIM, H., 1985. Interstitial Dorvilleidae (Polychaeta) from Europe, New Zealand and Australia. Zool. Scr. **14**, 183–199.

WESTHEIDE, W. & RAO, G. C., 1977. On some species of the genus *Hesionides* (Polychaeta, Hesionidae) from Indian sandy beaches. Cah. Biol. mar. **18**, 275–287.

WOLF, P. S., 1986. Four new genera of Dorvilleidae (Annelida: Polychaeta) from the Gulf of Mexico. Proc. Biol. Soc. Wash. **99**, 616–626.

Prof. Dr. Wilfried Westheide
Spezielle Zoologie, Fachbereich Biologie/Chemie, Universität Osnabrück
Postfach 44 69, D-4500 Osnabrück

Promesostoma teshirogii n. sp. (Plathelminthes, Rhabdocoela) aus Brackgewässern von Japan

Peter Ax

Inhaltsverzeichnis

Abstract	159
A. Einleitung	159
B. Beschreibung	161
C. Diskussion	162
Zusammenfassung	165
Literatur	165

Promesostoma teshirogii n. sp. (Plathelminthes, Rhabdocoela) from brackish water of Japan

Abstract

Promesostoma teshirogii n. sp. from brackish water areas of Honshu (Aomori Prefecture) is described. The new species is characterized by a stylet with a triangular excavation of the proximal part and with a funnel leading to the bursa seminalis. *P. teshirogii* is compared with *P. alaskana* Ax & Armonies, 1990 and *P. infundibulum* Ax, 1968.

A. Einleitung

Das Taxon *Promesostoma* v. Graff umfaßt heute ca. 30 gut charakterisierte Arten aus litoralen Meer- und Brackwassergebieten. Der letzten Zusammenstellung (EHLERS & SOPOTT-EHLERS 1989) ist *Promesostoma alaskana* Ax & Armonies, 1990 anzuschließen.

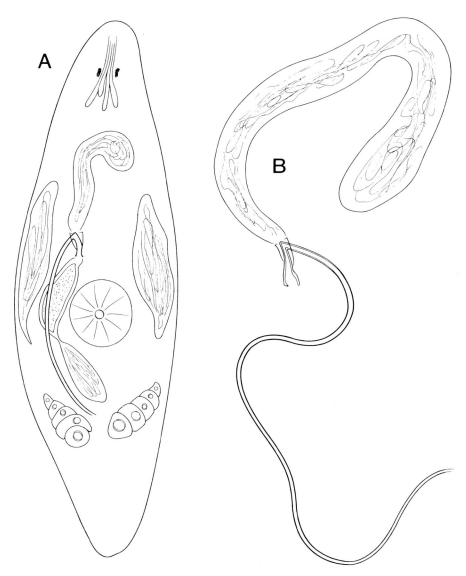

Abb. 1. *Promesostoma teshirogii*. A. Organisation. B. Stilettrohr mit Trichter und Bursa seminalis.

Promesostoma-Arten besiedeln ein weites Spektrum litoraler Lebensräume von lotischen Sandstränden bis zu lenitischen Salzwiesenarealen. Verschiedene Arten tolerieren niedrige Salinitäten und dringen bis in den Grenzbereich zum Süßwasser vor.

Bei Studien über die Plathelminthen-Besiedlung von Brackgewässern in Japan (Honshu, Aomori Prefecture) habe ich in Lagunen und Seen eine neue *Promesostoma*-Art gefunden. Ich widme sie Herrn Prof. Dr. Watru Teshirogi mit einem herzlichen Dank für die freundschaftliche Unterstützung meiner Arbeit an der Universität von Hirosaki.

B. Beschreibung

Fundorte: Japanische See Ostufer des Jusan Lake. Sand zwischen *Phragmites*. Salzgehalt: 15. 8. 90 = 5 ⁰/₀₀; 5. 9. 90 = Probenstelle unter Einfluß von Süßwasserzustrom.

Pazifischer Ozean Lagune an der Mündung des Takase River. Feinsand. Salzgehalt: 29. 8. 90 = 0–5 ⁰/₀₀

Takahoko Pond. Detritusreicher Sand der Uferzone. Salzgehalt: 29. 8. 90 = 3 ⁰/₀₀.

Material: Lebendbeobachtungen mit Zeichnungen und Mikrophotographien

Körperlänge ca. 0,7–0,8 mm, maximal 1 mm. Vorder- und Hinterende laufen konisch zu. Lebhaft schwimmende Art.

Mit Augen; ohne Pigmentierung. Pharynx in der Körpermitte. Paarige Hoden beginnen im zweiten Körperdrittel; sie verjüngen sich in Höhe des Pharynx. Die paarigen Germarien liegen im Hinterkörper.

Kopulationsorgan mit unpaarer Samenblase und ovoidem, muskulösen Bulbus. Das einfache Stilettrohr erreicht ca. 200 µm Länge. Von der proximal erweiterten Öffnung verläuft das Rohr zunächst ein Stück nach vorne, um dann mit scharfem Knick caudalwärts umzubiegen. In diesem Abschnitt erhält das Rohr die Form eines Dreiecks mit rostralwärts gerichteter Spitze (Abb. 2 A). Daneben wurden allerdings auch Individuen mit einem mehr rundlichen Knickbereich beobachtet (Abb. 2 B). Das distale Stilettende hat eine einfache Struktur; hier ist das Rohr schräg abgeschnitten; es kann leicht angeschwollen oder löffelförmig verdickt sein.

Dem proximalen Dreieck des Rohres sitzt ein besonderer Bursatrichter auf. Diese Struktur beginnt an einer leistenförmigen Verdickung des Rohres im Anschluß an einen kräftigen Sphinkter. Der schwach verfestigte Trichter läuft mit geschwungenen Wänden nach vorne und erweitert sich dabei leicht vasenförmig. Der kleine Trichter ragt aber kaum über den Knick des Rohres hinaus.

Der Trichter führt in eine schlauchförmige, nach vorne gerichtete Bursa seminalis.

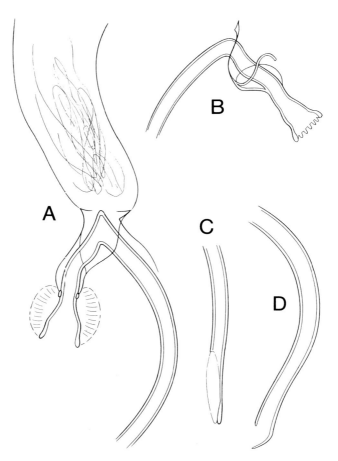

Abb. 2. *Promesostoma teshirogii*. A. Proximalende des Stilettrohres mit scharfem Knick und Bursatrichter. B. Proximalende des Stilettrohres eines anderen Individuums mit rundlichem Knickbereich. C. und D. Distalende des Rohres von zwei Individuen.

C. Diskussion

In der Struktur des Stilettrohres existieren gute Übereinstimmungen zwischen *Promesostoma teshirogii* n. sp. aus Japan und *Promesostoma alaskana* Ax & Armonies, 1990 von Alaska. Das betrifft die Länge und Form des Rohres mit dem proximalen Dreieck und der einfachen distalen Öffnung.

P. alaskana besitzt aber keinen Bursatrichter am proximalen Rohrende. In diesem Merkmal ist *P. teshirogii* vielmehr mit *P. infundibulum* Ax, 1968 ver-

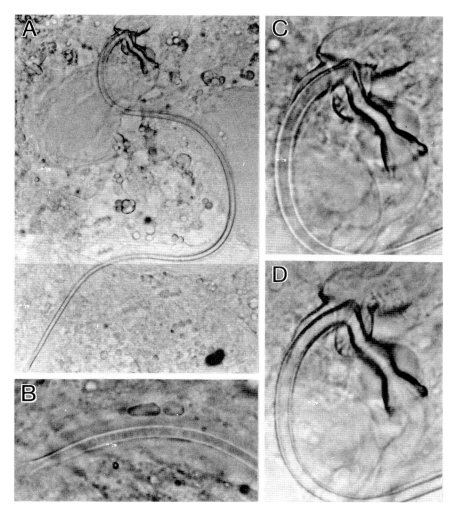

Abb. 3. *Promesostoma teshirogii*. Mikrophotographien. A. Stilettrohr. B. Distalende des Rohres. C. und D. Proximalende des Rohres mit scharfem Knick und Bursatrichter.

gleichbar. Der Trichter von *P. infundibulum* ist allerdings erheblich größer; er reicht weit über das Stilettrohr hinaus. Das Rohr selbst ist bei *P. infundibulum* in 2–3 Windungen aufgerollt (Ax 1968).

Mit diesem Vergleich sind klare Unterschiede zu den beiden Arten *P. alaskana* und *P. infundibulum* aufgezeigt. Eine Hypothese über die Schwesterart von *Promesostoma teshirogii* ist im Stand der Untersuchungen nicht möglich.

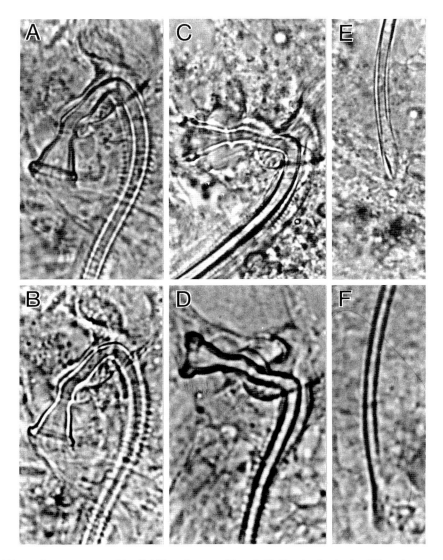

Abb. 4. *Promesostoma teshirogii.* Mikrophotographien. A–D. Proximalende des Stilettrohres verschiedener Individuen. E–F. Distalende.

Zusammenfassung

Promesostoma teshirogii n. sp. aus Brackgewässern von Honshu (Aomori Prefecture) wird beschrieben. Die neue Art ist charakterisiert durch eine dreieckige Struktur im proximalen Abschnitt des Stiletts und einen Trichter, der in die Bursa seminalis führt. *P. teshirogii* wird mit *P. alaskana* Ax & Armonies, 1990 und *P. infundibulum* Ax, 1968 verglichen.

Literatur

Ax P. (1968): Turbellarien der Gattung *Promesostoma* von der nordamerikanischen Pazifikküste. Helgoländer Wiss. Meeresunters. **18**, 116–123.

Ax, P. & W. Armonies (1990): Brackish water Plathelminthes from Alaska as evidence for the existence of a boreal brackish water community with circumpolar distribution. Microfauna Marina **6**, 7–109.

Ehlers, U. & B. Sopott-Ehlers (1989): Drei neue interstitielle Rhabdocoela (Plathelminthes) von der französischen Atlantikküste. Microfauna Marina **5**, 207–218.

Professor Dr. Peter Ax
II. Zoologisches Institut und Museum der Universität Göttingen
Berliner Str. 28, D–3400 Göttingen

The fastening of egg capsules of *Multipeniata* Nasonov, 1927 (Prolecithophora, Plathelminthes) on bivalves – an adaptation to living conditions in soft bottom

Peter Ax and Andreas Schmidt-Rhaesa

Contents

Abstract	167
A. Introduction	167
B. Material	169
C. Results	171
D. Discussion	171
Zusammenfassung	174
References	174

Abstract

Multipeniata Nasonov from Jusan Lake (Japan, Aomori Prefecture) fastens its egg capsules to the shell of living *Corbicula japonica* (Bivalvia). The capsules are mainly deposited at the hinge region of the shell. This phenomenon is interpreted as an evolutionary adaptation to the invasion of a soft bottom habitat.

A. Introduction

Solid capsules for the acceptance of ova and yolk cells belong to the ground pattern of the Neoophora, a community of descent within the Plathelminthes which have ectolecithal eggs (EHLERS 1985).

The capsule structures from many species of the Proseriata and Kalyptorhynchia were examined by our research group. These were mainly capsules that were deposited at the bottom of Petri dishes (GIESA 1966, 1968; HOXHOLD 1971; SOPOTT 1973).

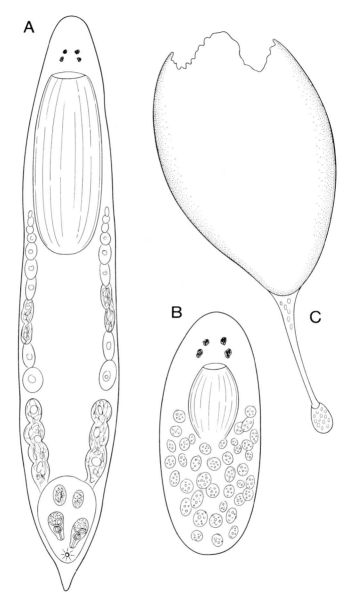

Fig. 1: *Multipeniata* spec. (Honshu, Aomori Prefecture). A. Adult with several copulatory organs at the posterior end, showing intense red colouration. Length 3–4 mm (Kominato River).
B. Juvenile, forced out of the capsule, 350–400 μm long (Jusan Lake).
C. Empty capsule with irregular opening. Capsule length ca. 350 μm, stem ca. 125 μm (Jusan Lake).

However, the fastening of Neoophora egg capsules on living organisms has been rarely observed. Besides an epizoic bond exhibited by *Plagiostomum oyense* (Prolecithophora) and species of *Idotea* (Isopoda) (BEAUCHAMP 1921; NAYLOR 1952, 1955), different molluscs may serve as biotic substrate. Species of *Monocelis* (Proseriata) attach egg capsules to bivalves (*Donax, Tellina*) (GIARD 1897) and gastropods (*Hydrobia, Littorina*) (GIESA 1966); an *Archilopsis*-species fastens the capsules deep in the sediment to the shell of *Macoma balthica* (REISE 1985, p. 121).

During studies on Plathelminthes from brackish waters in northern Japan (Honshu, Aomori Prefecture) regularly stemmed egg capsules of *Multipeniata* were found on young individuals of the bivalve *Corbicula japonica* Prime (Corbiculidae). With the description of this matter we connect an evolutionary and ecological interpretation.

The Department of Biology of the Hirosaki University was the starting point for the research. The senior author gives many thanks to Prof. Dr. W. Teshirogi, Prof. Dr. S. Ishida and Mrs. K. Ohkawa for their untiring support in august and september 1990.

B. Material

The taxon *Multipeniata* was described with the two species *M. kho* Nasonov, 1927 and *M. batalansae* Nasonov, 1927 from brackish river mouths on the Russian coast of the Sea of Japan (NASONOV 1927, 1932).

On Honshu (Aomori Prefecture), *Multipeniata* settles in diverse brackish water biotopes at the Sea of Japan and the Pacific Ocean. The questionable differentiation of two species made by NASONOV and connected with this the problem of the species identity of northern Japanese populations of *Multipeniata* are discussed elsewhere.

Observations of the egg capsules were made along the east shore of Jusan Lake which has low salinity (between 10 and 0 $^o/_{oo}$ according to measurements made on August, 8 and September, 9, 1990). Young individuals of *Corbicula japonica* densely colonize the soft sediment along this shore.

Identification of egg capsules as belonging to *Multipeniata* was made by examinations of newly emerged juveniles. Individuals ready to hatch possess 4 eyes like the adult (Fig. 1,2). This is not the case with any other plathelminth species within the research area. In addition to the research in Japan, 12 fixed bivalves with capsules were examined at Göttingen.

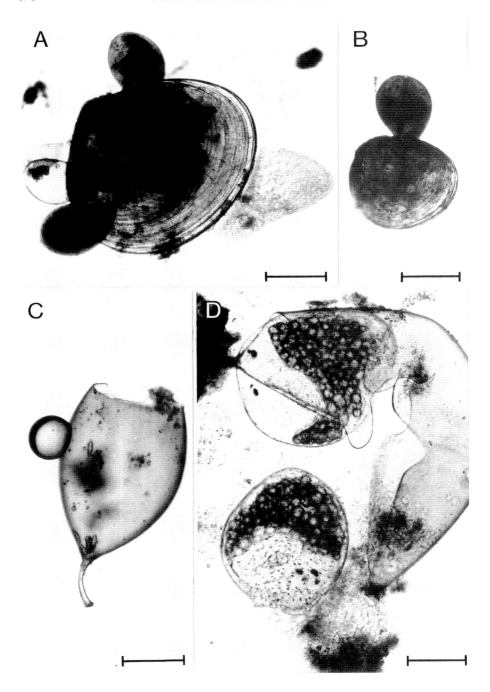

C. Results

The observations about egg capsules of *Multipeniata* are a byproduct of ecological and geographical research on Plathelminthes from brackish waters of northern Japan. Although quantitative measurements aren't available, the very regular supply of young bivalves with capsules when brought with the sediment into Petri dishes must be emphasized as a striking phenomenon. Normally 1–3 capsules are fixed to one bivalve; at maximum 11 capsules were observed. The localisation of the capsules is of special interest. They are predominantly fixed in the hinge region of the shell. One capsule may contain up to 3 embryos.

The lengths of the capsules range between 200 and 350 µm. Empty capsules are minimal about 100 µm in length. The stem is 60–125 µm long; it is fastened to the shell with a broad plate (70 µm diameter).

The egg capsules of *Multipeniata* do not possess a preformed operculum with seam. The openings of empty capsules have irregular bursted edges. The structure differs from capsule to capsule (Fig. 4 A,B). Some capsules have a regular edge around their opening (Fig. 2 C). Capsules of this type may belong to another species of the Neoophora.

D. Discussion

At present there exist no established idea about lifestyle and biotope bond of the stem species of the Neoophora. However, it may be assumed that egg capsules are primarily fastened to abiotic solid substrates that are available in most environments except soft bottom. This hypothesis is more probable than the assumption of a primary linkage of the capsules to biotic substrates like mollusc shells or exoskeletons of arthropods – substrates that probably were not available during the evolution of the egg capsules of plathelminths.

From this point of view the fastening of capsules of *Multipeniata* can be regarded as a "strategy" to conquer a muddy biotope as a new habitat. The fastening to pebbles or other abiotic materials in the mud can not be advantageous; in the case of a shift into deeper sediments the lack of oxygen would stop the development of the embryos within the capsules. The bivalve *Corbicula*

◀ Fig. 2: Light-microscope photographs. A. 4 egg capsules of *Multipeniata* on living young *Corbicula japonica* (Bivalvia). B. Egg capsule on *Corbicula japonica* with two embryos. C. Small capsule of unknown origin with a regular edge along the opening. D. Two young individuals of *Multipeniata* forced from an egg capsule (Jusan Lake). Scale: A,B = 100 µm; C,D = 50 µm.

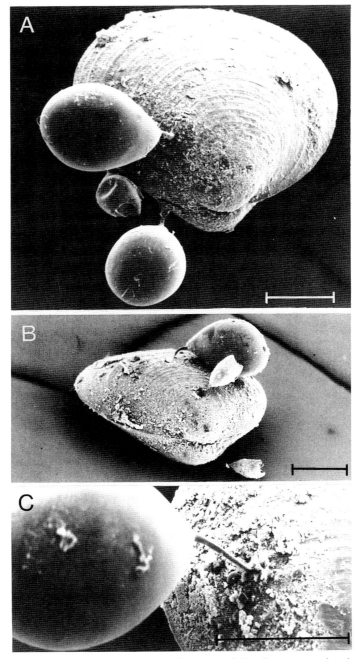

Fig. 3: Scanning Electron Microscope photographs. A. Two full and one empty, shrunken capsules of *Multipeniata* on *Corbicula japonica*. B. One full and two empty capsules. C. Capsule with stem (Jusan Lake). Scale: A,B,C = 100 μm.

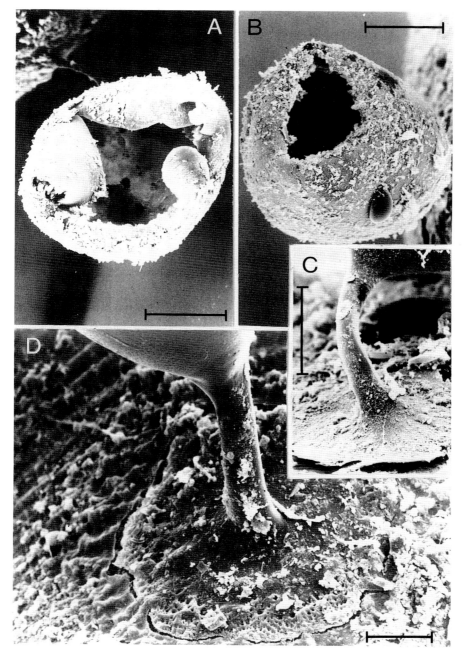

Fig. 4: Scanning Electron Microscope photographs. A,B. Two empty capsules from Fig. 3. B of *Multipeniata* with irregular openings. C,D. Stem of the capsule with broad basic plate on the bivalve shell (Jusan Lake). Scale: A,B,C = 20 μm; D = 10 μm.

japonica on the other hand settles constantly at the border between water and sediment, and digs its way up to the surface if it becomes buried. *Corbicula japonica* orientates the hinge region of its shell towards the water and this is precisely the region *Multipeniata* chooses for the fastening of the egg capsules. This behaviour can be interpreted as an adaptation to living conditions in soft bottom.

Our observations raise new questions. It remains to be seen if the deposition of *Multipeniata* egg capsules on *Corbicula japonica* is compulsory or optional. How much does mechanical or chemical signals of the bivalve induce the deposition? This study may also stimulate a comparative analysis of the deposition of plathelminth eggs in soft bottom.

Zusammenfassung

Multipeniata Nasonov aus dem Jusan Lake (Japan, Aomori Prefecture) befestigt gestielte Eikapseln an der Schale lebender *Corbicula japonica* (Bivalvia). Die Kapseln werden insbesondere in der Schloßregion der Muschel abgesetzt. Das Phänomen wird als eine evolutive Adaption bei der Invasion in den Lebensraum des Weichbodens interpretiert.

References

BEAUCHAMP, P. DE (1921) Sur un nouveau *Plagiostomum* et ses rapports avec un Isopode. Bull. Soc. Zool. de France **46**, 169–176.
EHLERS, U. (1985): Das Phylogenetische System der Plathelminthes. G. Fischer, Stuttgart, New York, 317 pp.
GIARD, A. (1897): Sur la ponte des Rhabdocoels de la famille des Monotidae. Compt. rend. Soc. Biol. Paris **4**.
GIESA, S. (1966): Die Embryonalentwicklung von *Monocelis fusca*. Z. Morph. Ökol. Tiere **57**, 137–230.
— (1968): Die Eikapseln der Proseriaten (Turbellaria, Neoophora). Z. Morph. Ökol. Tiere **61**, 338–346.
HOXHOLD, S. (1971): Eigebilde interstitieller Kalyptorhynchia von der deutschen Nordseeküste. Mikrofauna des Meeresbodens **7**, 253–293.
NASONOV, N. (1927): Über eine neue Familie Multipeniatidae (Alloeocoela) aus dem japanischen Meer mit einem aberranten Bau der Fortpflanzungsorgane. Bulletin de l'Academie des Sciences de l'URSS 1927, 865–874.
— (1932): Zur Morphologie der Turbellaria Rhabdocoelida des japanischen Meeres. Trav. Lab. Zool. Exper. et Morph. Animaux **2**, 1–112.
NAYLOR, E. (1952): On *Plagiostomum oyense* de Beauchamp, an epizoic Turbellarian new to the british fauna. Rep. Port Erin Mar. Biol. Stat. **64**, 15–21.
— (1955): The seasonal abundance on *Idotea* of the cocoons of the flatworm *Plagiostomum oyense* de Beauchamp. Rep. Port Erin Mar. Biol. Stat. **67**, 1–6.

REISE, K. (1985): Tidal flat ecology. An experimental approach to species interactions. Springer, Berlin. 191 pp.
SOPOTT, B. (1973): Jahreszeitliche Verteilung und Lebenszyklen der Proseriata (Turbellaria) eines Sandstrandes der Nordseeinsel Sylt. Mikrofauna des Meeresbodens **15**, 255–358.

Professor Dr. Peter Ax and *Andreas Schmidt-Rhaesa,*
II. Zoologisches Institut der Universität Göttingen
Berliner Straße 28, D-3400 Göttingen

Coelogynopora sequana nov. spec. (Proseriata, Plathelminthes) aus der Seine-Mündung

Beate Sopott-Ehlers

Inhaltsverzeichnis

Abstract	177
A. Einleitung	178
B. Ergebnisse	178
C. Diskussion	181
Zusammenfassung	183
Abkürzungen in den Abbildungen	183
Literatur	184

Coelogynopora sequana nov. spec. (Proseriata, Plathelminthes) from the Seine estuary

Abstract

Coelogynopora sequana nov. spec. is described from sandy sediments from the estuary of the river Seine.

The animals found by KARLING (1958, 1974) in the Baltic and ascribed to the unsufficiently known species *"Coelogynopora bresslaui"* Steinböck, 1924 are identical with the new species.

STEINBÖCK (1924) did not provide detailed informations on the hard structures in the male copulatory organ, which allow an exact identification of the species from living material. This means, *"C. bresslaui"* has to be considered a *nomen dubium*.

The new species *C. sequana* and *C. steinböcki*, *C. hamulis* and *C. juxtaforcipis* are a monophyletic species-group within the taxon *Coelogynopora*.

A. Einleitung

Obwohl die interstitielle Fauna der europäischen Küsten seit mehreren Jahrzehnten untersucht wird, finden sich immer wieder neue, unbeschriebene Arten.

In unregelmäßigen Abständen wurden während der letzten Jahre Sandproben entlang der französischen Küsten genommen. Dabei konnten mehrere Species entdeckt werden, die neu für die Wissenschaft waren (SOPOTT-EHLERS 1976; EHLERS 1980; EHLERS & EHLERS 1980; EHLERS & SOPOTT-EHLERS 1989).

In der vorliegenden Studie wird ein Vertreter der Coelogynoporidae beschrieben, der bereits von KARLING bei Falsterbo, Schweden beobachtet wurde. KARLING (1958, 1974) schrieb seine Funde der nur ungenügend bekannten Art „*Coelogynopora bresslaui*" Steinböck, 1924 zu. Für eine sichere Identifikation dieser Art wäre die genaue Kenntnis der Hartstrukturen des männlichen Begattungsorgans nach Lebendbeobachtungen erforderlich; diese Informationen liegen jedoch von STEINBÖCK (1924) nicht vor. „*C. bresslaui*" ist somit als ein *nomen dubium* zu betrachten. Für das nach Lebendbeobachtungen bekannte Material von KARLING (1958, 1974) und von mir wird daher die neue Art *C. sequana* errichtet.

B. Ergebnisse

Coelogynopora sequana nov. spec.

(syn. „*Coelogynopora bresslaui*" *sensu* KARLING 1958, 1966, 1974).

Fundorte: 1. Französische Kanalküste bei Mont St. Michel; 1 Exemplar; September 1983. 2. Seine-Mündung bei Honfleur in detritushaltigen Sedimenten (Locus typicus); 11 Exemplare; September 1990.

Weitere Verbreitung: Schwedische und finnische Ostseeküste bei Falsterbo (KARLING 1958) bzw. bei Hangö (KARLING 1974).

Material: Lebendbeobachtungen, Dauerpräparate.

Etymologie: Der Artname leitet sich vom lateinischen Namen der Seine ab.

Die Individuen erreichen im gestreckten Zustand eine Länge von 5,5–6 mm. Das Rostralende trägt frontal einen dichten Saum kurzer Tasthärchen. Lateral stehen einzelne Tastborsten. Wenig prominente Haftorgane sind am ganzen Körper vorhanden; am Schwanzende treten sie konzentriert auf. Ballonförmige, lichtbrechende Hautdrüsen (Abb. 1 C), Paracniden, mit einem Durchmesser von etwa 10 µm verteilen sich über den gesamten Körper. Bei stark gequetschten Tieren nehmen die Paracniden eine flaschenförmige Gestalt an. In der Region des Vorderendes treten deutlich die Kanäle des Protonephridialsystems hervor.

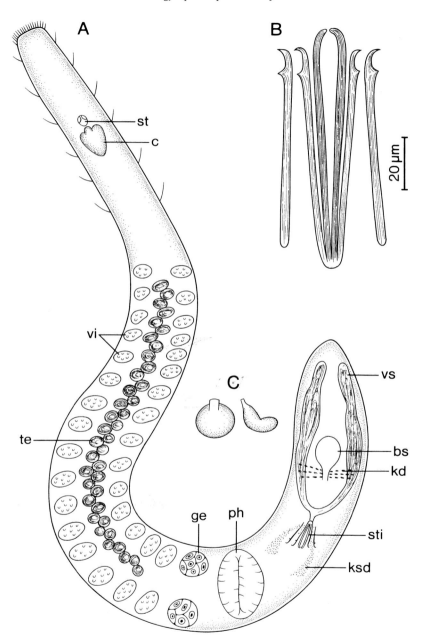

Abb. 1. A. Organisationsschema. B. Stilettapparatur. C. Hautdrüsen (Paracniden).

Das Gehirn erscheint dreilappig. Der dorsoventral verlaufende Pharynx befindet sich am Beginn des hinteren Körperdrittels.

Männliche Geschlechtsorgane. Die männlichen Gonaden zeigen die für die Coelogynoporiden typische Anordnung. Die Hodenfollikel bilden einen unpaaren medianen Strang in der praepharyngealen Region.

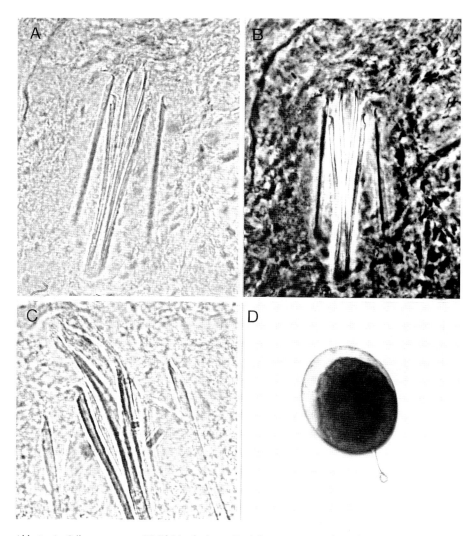

Abb. 2. A. Stilettapparatur, Hellfeldaufnahme. B. Stilettapparatur, Phasenkontrastaufnahme. C. Spitzen der medianen Nadeln des Zentralbündels in dorsaler Aufsicht. D. Eikapsel.

Die Stilettapparatur (Abb. 1 B, 2 A–C) des männlichen Begattungsorgans besteht aus einem zentralen Bündel basal zu einer Platte miteinander verwachsener Nadeln und zwei unmittelbar neben diesem Bündel liegenden Begleitnadeln.

Das zentrale Nadelbündel setzt sich aus 4 Elementen zusammen. In der Mitte stehen zwei schlanke Nadeln (Länge: 85–90 µm) mit leicht gebogener Spitze. Diese Spitze erscheint bei ventraler oder dorsaler Aufsicht blattförmig verbreitert (Abb. 2 C). Neben diesen medianen Nadeln tritt ein laterales Nadelpaar auf. Diese lateralen Nadeln tragen unterhalb ihrer gebogenen Spitze einen Hakenfortsatz (Sporn) und messen 75–78 µm.

Auch die freistehenden Begleitnadeln haben unterhalb der Spitze einen Hakenfortsatz. Die Begleitnadeln erreichen eine Länge von 60–63 µm.

Kornsekretdrüsen ziehen rechts und links des Zentralbündels frontad. Ein Kornsekretreservoir ist nicht ausgebildet. Stränge des Kornsekrets legen sich dicht an die Nadeln des Zentralbündels an. Zwei langgestreckte Samenblasen ziehen bis zur Schwanzspitze. Unmittelbar vor der Stilettapparatur vereinigen sie sich zu einem kurzen Ductus communis.

Weibliche Geschlechtsorgane. Die Vitellarien sind als Einzelfollikel in serialer Anordnung ausgebildet. Sie beginnen in einiger Entfernung hinter dem Gehirn und enden in der praepharyngealen Region dicht vor den Germarien. An keinem der untersuchten Individuen traten postpharyngeal Vitellarfollikel auf. Paarige, kompakte Germarien liegen vor der Pharynxwurzel. Sie enthalten mehrere Germocyten, deren Größe caudalwärts zunimmt. Zwischen den Samenblasen ist ein kleines Bursalorgan ausgebildet. In seinen frontalen halsförmigen Abschnitt münden Drüsen, deren Sekret grobkörnig ist.

Eikapseln (Abb. 2 D). Einige der in Petrischalen gehälterten Tiere legten Eikapseln ab. Diese haben eine länglichovale Form und einen Durchmesser von 209 × 171 µm. Die Eikapseln sind mit einem kurzen Stiel versehen (Länge 47–48 µm), dessen Ende knotenförmig verdickt ist. Eine Ausschlüpföffnung (Deckel) konnte nicht beobachtet werden. Die Kapseln erscheinen im Durchlicht hellgelb.

C. Diskussion

Coelogynopora-Species, deren Stilettapparatur basal zu einer Platte verwachsene Nadeln enthält, wurden bisher von STEINBÖCK 1924 ("*C. bresslaui*"), MEIXNER 1938 (*C. tenuis*), TAJIKA 1978 (*C. birostrata*), AX & SOPOTT-EHLERS 1979 (*C. nodosa*), SOPOTT-EHLERS 1980 (*C. steinböcki, C. hamulis*) und RISER 1981 (*C. juxtaforcipis*) beschrieben.

C. tenuis und *C. birostrata* scheiden für einen näheren Vergleich mit *C. se-*

quana aus, da das zentrale Nadelbündel caudad gerichtet ist. Auch *C. nodosa* kommt nicht in Betracht, da bei dieser Art zwei Komplexe verwachsener Nadeln auftreten.

Die Stilettapparatur der hier dargestellten Art stimmt in Anzahl, Form und Länge der Nadeln mit den Gegebenheiten bei den von KARLING (1958, 1974) als „*C. bresslaui*" identifizierten Individuen überein.

Wie KARLING bereits 1958 erwähnt, erfolgte die Originalbeschreibung von „*C. bresslaui*" lediglich nach Schnittserien. Dies macht eine sichere Identifizierung unmöglich, da die Hartstrukturen der Stilettapparatur in Schnitten gewöhnlich splittern und feinere Details der Stilette nur im Quetschpräparat sicher zu erkennen sind. Gerade diese Hartstrukturen aber liefern das wesentliche Merkmal zur Artidentifizierung. „Die Rekonstruktion eines verwickelten Kutikularapparates nach Schnitten leitet aber selten zu einem befriedigenden Resultat" (KARLING 1958, p. 564).

STEINBÖCK (1924, p. 464, fig. 3 b) zeichnet für „*C. bresslaui*" ein zentrales Nadelbündel bestehend aus 8 Elementen. In der Sagittalrekonstruktion (STEINBÖCK l. c., p. 462, fig. 1; MEIXNER 1938, p. 59, fig. 57) bildet STEINBÖCK 4 Nadeln ab, zwei davon mit Hakenfortsatz. Da man davon ausgehen muß, daß im idealen Sagittalschnitt nur die Hälfte der Nadeln des zentralen Bündels getroffen wird, weist auch die Sagittalrekonstruktion auf die Existenz von mehr als 4 Nadeln im Zentralbündel hin.

Nach unserem gegenwärtigen Wissensstand könnten STEINBÖCK auch die ebenfalls aus der Nordsee bekannte Art *C. hamulis*, eventuell sogar die von der amerikanischen Atlantikküste beschriebene *C. juxtaforcipis* vorgelegen haben. Die von Helgoland stammende Species *C. steinböcki* scheidet dagegen aus, da keine ihrer Zentralnadeln einen Hakenfortsatz (Sporn) trägt.

Da es nach den vorliegenden Informationen von STEINBÖCK unmöglich ist zu sagen, wie die Stilettapparatur von „*C. bresslaui*" gestaltet ist und so lebende Individuen als Vertreter dieser Art zu identifizieren, muß „*C. bresslaui*" gemäß den internationalen Regeln für Nomenklatur (KRAUS 1970, p. 78; MAYR 1975, p. 347; INTERNATIONAL COMMISSION ON ZOOLOGICAL NOMENCLATURE 1985, p. 260) als *nomen dubium* betrachtet werden.

Die von KARLING (1958, 1974) als „*Coelogynopora bresslaui*" dargestellten Tiere identifiziere ich als *C. sequana*.

Aus dem Atlantik und seinen Nebenmeeren sind derzeit folgende Vertreter des Taxons *Coelogynopora* mit einem unpaaren, basal zu einer Platte verwachsenen und rostrad gerichteten Bündel von Zentralnadeln sicher bekannt: *C. steinböcki* Sopott-Ehlers, 1980, *C. hamulis* Sopott-Ehlers, 1980, *C. juxtaforcipis* Riser, 1981 und die neue Art *C. sequana*. In Form, Anzahl und Länge der Nadeln des Zentralbündels sowie der Position der Begleitnadeln setzt sich

C. *sequana* deutlich gegen C. *hamulis* und C. *juxtaforcipis* ab. Darüber hinaus unterscheidet sich C. *sequana* noch durch die großen Paracniden und den Mangel eines Kornsekretreservoirs von C. *hamulis*.

Die Position der Begleitnadeln und das Fehlen einer Vesicula granulorum hat C. *sequana* mit C. *steinböcki* gemein. Jedoch setzt sich das zentrale Nadelbündel von C. *steinböcki* aus insgesamt 8 Nadeln gegenüber 4 Nadeln bei C. *sequana* zusammen. Ferner fehlt sämtlichen Nadeln von C. *steinböcki* ein Hakenfortsatz. Darüber hinaus sind deutliche Unterschiede durch die Gestalt der Hautdrüsen gegeben. C. *steinböcki* besitzt dicht an dicht stehende Hautdrüsen wie sie beispielsweise auch bei C. *gynocotyla* Steinböck, 1924 (cf. SOPOTT 1972, p. 20, fig. 5), C. *biarmata* Steinböck, 1924 und bei *Cirrifera*-Species auftreten, während bei C. *sequana* einzeln gelegene Paracniden differenziert sind.

Innerhalb des Taxons *Coelogynopora* stellen C. *steinböcki*, C. *hamulis*, C. *juxtaforcipis* und C. *sequana* ein monophyletisches Teiltaxon dar mit der Autapomorphie „Zentrales Nadelbündel unpaar, basal verwachsen und rostrad gerichtet."

Zusammenfassung

Aus sandigen, detritushaltigen Sedimenten der Seine-Mündung wird *Coelogynopora sequana* nov. spec. beschrieben.

Die von KARLING (1958, 1974) an Ostseestränden beobachteten und der ungenügend bekannten Art „*Coelogynopora bresslaui*" Steinböck, 1924 zugeschriebenen Individuen sind mit der neuen Species identisch.

STEINBÖCK (1924) gibt für „C. *bresslaui*" keine genauen Informationen über die Hartstrukturen des männlichen Begattungsorgans, die eine sichere Identifikation dieser Art nach Lebendbeobachtungen ermöglichen würden. „C. *bresslaui*" ist somit als ein *nomen dubium* zu betrachten.

Die neue Species C. *sequana* bildet zusammen mit C. *steinböcki*, C. *hamulis* und C. *juxtaforcipis* ein monophyletisches Teiltaxon innerhalb des Taxons *Coelogynopora*.

Abkürzungen in den Abbildungen

bs	Bursa	ksd	Kornsekretdrüsen	te	Hodenfollikel
c	Gehirn	ph	Pharynx	vi	Vitellarfollikel
ge	Germarium	st	Statocyste	vs	Vesicula seminalis
kd	Kittdrüsen	sti	Stilettnadeln		

Literatur

Ax, P. & B. Sopott-Ehlers (1979): Turbellaria Proseriata von der Pazifikküste der USA (Washington). II. Coelogynoporidae. Zool. Scr. **8**, 25–35.

Ehlers, B. & U. Ehlers (1980): Zur Systematik und geographischen Verbreitung interstitieller Turbellarien der Kanarischen Inseln. Mikrofauna Meeresboden **80**, 1–23.

Ehlers, U. (1980): Interstitial Typhloplanoida (Turbellaria) from the area of Roscoff. Cah. Biol. Mar. **21**, 155–167.

Ehlers, U. & B. Sopott-Ehlers (1989): Drei neue interstitielle Rhabdocoela (Plathelminthes) von der französischen Atlantikküste. Microfauna Marina **5**, 207–218.

International Commission On Zoological Nomenclature (1985): International Code of Zoological Nomenclature, 3. Aufl. 338 pp., University of California Press, Berkeley, Los Angeles.

Karling, T. G. (1958): Zur Kenntnis der Gattung *Coelogynopora* Steinböck (Turbellaria Proseriata). Ark. Zool. Ser. 2, **11**, 559–567.

– (1966): Marine Turbellaria from the Pacific coast of North America. IV. Coelogynoporidae and Monocelididae. Ark. Zool.. Ser. 2, **18**, 493–528.

– (1974): Turbellarian fauna of the Baltic proper. Identification, ecology and biogeography. Fauna Fennica **27**, 1–101.

Kraus, O. (ed.) (1970): Internationale Regeln für die zoologische Nomenklatur. Beschlossen vom XV. Internationalen Kongreß für Zoologie. 2. Aufl., 92 pp., Verlag Waldemar Kramer, Frankfurt/M.

Mayr, E. (1975): Grundlagen der zoologischen Systematik. 370 pp., Verlag Paul Parey, Hamburg, Berlin.

Meixner, J. (1938): Turbellaria (Strudelwürmer) I. (Allgemeiner Teil). Die Tierwelt der Nord- und Ostsee, **IV** b, 1–146.

Riser, N. (1981): New England Coelogynoporidae. Hydrobiologia **84**, 139–145.

Sopott, B. (1972): Systematik und Ökologie von Proseriaten (Turbellaria) der deutschen Nordseeküste. Mikrofauna Meeresboden **13**, 1–72.

Sopott-Ehlers, B. (1976): Interstitielle Macrostomida und Proseriata (Turbellaria) von der französischen Atlantikküste und den Kanarischen Inseln. Mikrofauna Meeresboden **60**, 1–35.

– (1980): Zwei neue *Coelogynopora*-Arten (Turbellaria, Proseriata) aus dem marinen Sublitoral. Zool. Scr. **9**, 161–163.

Steinböck, O. (1924): Untersuchungen über die Geschlechtstrakt-Darmverbindung bei Turbellarien nebst einem Beitrag zur Morphologie des Trikladendarmes. Z. Morph. Ökol. Tiere **2**, 461–504.

Tajika, K. J. (1978): Zwei neue Arten der Gattung *Coelogynopora* Steinböck, 1924 (Turbellaria, Proseriata) aus Hokkaido, Japan. J. Fac. Sci. Hokkaido Univ. (6) 21, 295–316.

Dr. Beate Sopott-Ehlers
II. Zoologisches Institut und Museum der Universität Göttingen
Berliner Str. 28, D-3400 Göttingen

Coelogynopora faenofurca nov. spec. (Proseriata, Plathelminthes) aus Wohnröhren des Polychaeten *Arenicola marina*

Beate Sopott-Ehlers

Inhaltsverzeichnis

Abstract 185
A. Einleitung 185
B. Material und Methode 186
C. Artbeschreibung 186
Zusammenfassung 190
Abkürzungen in den Abbildungen 190
Literatur 190

Coelogynopora faenofurca nov. spec. from the burrows of the polychaete *Arenicola marina*

Abstract

Coelogynopora faenofurca n. sp. is described from the German North sea coast. This species is most closely related to *Coelogynopora erotica* inhabiting sand beaches at the American Atlantic coast.

C. faenofurca has been exclusively found in the burrows of *Arenicola marina*.

A. Einleitung

Eingehende Untersuchungen der Wohnröhren des Polychaeten *Arenicola marina* haben gezeigt, daß das Sediment dieser Bauten eine arten- und individuenreiche Sandlückenfauna beherbergt (REISE & AX 1979; REISE 1984; SCHERER 1985).

Die Vertreter der freilebenden Plathelminthes stellen mit 83 Arten die größte Gruppe dieser Mikrofauna (SCHERER l. c.). Einige dieser Species sind neu für die Wissenschaft.

In dem vorliegenden Beitrag wird ein Vertreter der Proseriata beschrieben.

Herrn cand. rer. nat. W. Ahlrichs danke ich für die Beschaffung des Substrates.

B. Material und Methode

Das Sediment wurde Anfang Januar 1991 aus dem Sandwatt des Königshafens der Nordseeinsel Sylt entnommen.

Die Substratproben wurden anschließend 14 Tage bei einer konstanten Temperatur von 12° C aufbewahrt, bevor die Extraktion der Tiere mit der Seewassereis-Methode nach UHLIG erfolgte. Auf diese Weise konnten bereits früh im Jahr geschlechtsreife Individuen gewonnen werden.

C. Artbeschreibung

Coelogynopora faenofurca nov. spec.

(Abb. 1–3)

Coelogynopora faenofurca n. n. in REISE & AX (1979, p. 233 Tab. 2).

Fundort: Deutsche Nordseeküste. Sylt. Im Sandwatt des Königshafens (Locus typicus). 17 Exemplare. Januar 1991.

Material: Lebendbeobachtungen, Dauerpräparate.

Simultangeschlechtliche Individuen messen 6–7 mm. Die Kopfspitze ist mit einem Saum kurzer Tasthärchen versehen. An den Seiten des Rostralendes stehen einige lange Tastborsten. Bereits an leicht gequetschten Tieren fallen bei mikroskopischer Betrachtung lichtbrechende Hautdrüsen, Paracniden, auf (Abb. 1 B). Diese haben einen Durchmesser von 19–25 µm und sind in der Aufsicht rund, in Seitenansicht amphorenförmig gestaltet.

Der Pharynx verläuft in dorsoventraler Richtung.

Die Statocyste (Abb. 3 B) zeigt einen ellipsoiden Statolithen. Deutlich sind zwei runde Zellkerne (Nebensteinchen) der akzessorischen Zellen zu sehen.

Männliche Geschlechtsorgane. Der unpaare Strang der Hodenfollikel durchzieht die beiden vorderen Körperdrittel. Zwei schlanke Samenblasen erstrecken sich zum Schwanzende. Die Vesiculae seminales vereinigen sich rostral in einem Ductus communis. Diesem schließt sich eine rundliche Vesicula granulorum an, in die die Kornsekretdrüsen ihr Sekret abgeben. Dem Kornsekretreservoir sitzt ein zungenförmiges Trichterrohr von 90–100 µm Länge auf. Dieses wird von zwei schlanken Hakennadeln (Länge: 75–80 µm) flankiert (Abb. 1 C; 2 A,B). Es folgt ein sehr kräftiges Nadelpaar von 70–80 µm Länge. Diese Nadeln können auf den ersten Blick wie zusätzliche Trichterrohre erscheinen. In den meisten Quetschpräparaten zeigt ihre „Ventralseite" nach oben. Man schaut auf gegabelte Hakenfortsätze. Das zentrale Nadelbündel der Stilettapparatur weist noch 3–4 Paar weitere schlanke, leicht geschwungene Hakennadeln (Länge: 65–75 µm) auf. Zwei kurze, 30–40 µm messende Nadeln flankieren das Zentralbündel. Sie sind schwach gebogen und tragen median einen kleinen Sporn. Auffällig sind Lage und Form der Begleitnadeln. Zwei Bündel (Abb. 1 C; 2 A) aus je 6–8 Einzelelementen sind rostral der zentralen Nadeln differenziert. Diese Begleitnadeln erreichen eine Länge von 50–55 µm. Sie haben die Form einer zweizinkigen Gabel und weisen mit ihrer Spitze caudalwärts.

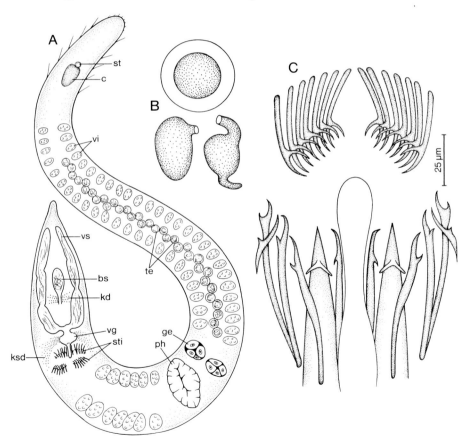

Abb. 1. A. Organisationsschema. B. Hautdrüsen unterschiedlich gequetscht. C. Stilettapparatur.

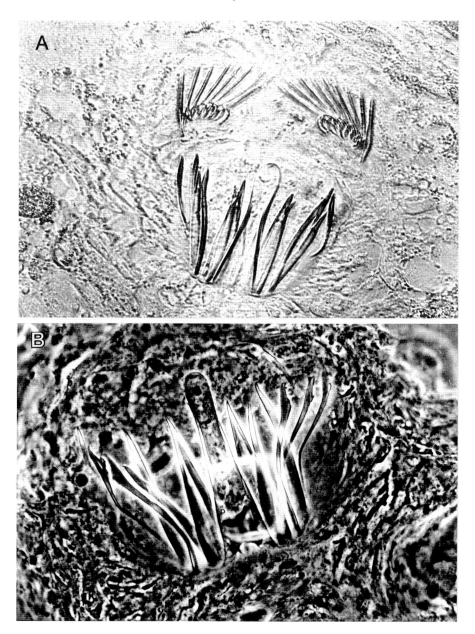

Abb. 2. A. Stilettapparatur. Unterschiedliche Nadelanzahl in den Bündeln der Begleitnadeln. (Hellfeldaufnahme). B. Nadeln des zentralen Bündels. (Phasenkontrastaufnahme).

Weibliche Geschlechtsorgane. Vitellarien und Germarien liegen in der für Coelogynoporiden charakteristischen Position. Zwischen den beiden Vesiculae seminales ist ein länglich ovales Bursalorgan differenziert. In seinen vorderen Abschnitt münden Drüsen, die grobscholliges Sekret sezernieren.

Eikapseln (Abb. 3 B). Die hellgelben Eikapseln sind von tropfenförmiger Gestalt und haben einen Durchmesser von circa 380 : 210 µm. Am spitz zulaufenden Pol setzt mit breiter Basis ein 190 µm langer Stiel an, der in einer knotenförmigen Verdickung (28 µm) endet.

Diskussion. Die Gestaltung der Stilettapparatur zeigt innerhalb der Coelogynoporidae eine erstaunliche Vielfalt. Ähnlich kompliziert gestaltete Hartstrukturen des männlichen Begattungsorgans mit einem Trichterrohr innerhalb des Zentralbündels und einer Vielfalt rostral lokalisierter Begleitnadeln sind auch bei der von der Nordsee- und der Atlantikküste bekannten Art *Coelogynopora forcipis* (SOPOTT-EHLERS 1976) und der von der französischen Atlantikküste beschriebenen Species *C. cassida* (SOPOTT-EHLERS 1985) vorhanden. Beide Arten kommen jedoch für einen näheren Vergleich mit *C. faenofurca* allein aufgrund der Gestaltung des Komplexes der Begleitnadeln nicht in Betracht.

Weitgehende Übereinstimmungen bestehen dagegen mit der von der amerikanischen Atlantikküste stammenden *C. erotica* Riser (RISER 1981). Eine Artidentität zwischen der neuen Species und *C. erotica* läßt sich jedoch durch einen

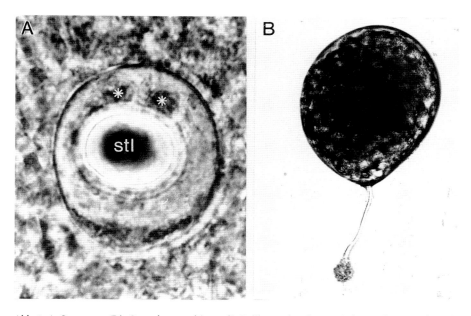

Abb. 3. A. Statocyste. Die Sternchen markieren die Zellkerne der akzessorischen Zellen. B. Eikapsel.

Vergleich der Anzahl und der Maße der Begleitnadeln (*C. faenofurca*: 6–8 Nadeln pro Bündel mit einer Länge von 50–55 µm; *C. erotica*: 12 Nadeln pro Bündel mit einer Länge von 60–90 µm) ausschließen. Hinsichtlich der zentralen Stilettapparatur scheinen keine so auffälligen Unterschiede zwischen den beiden Arten zu bestehen. Die Nadeln des Zentralbündels von *C. erotica* könnten nach der knappen Beschreibung von RISER (1981, p. 141) zu urteilen jenen von *C. faenofurca* weitgehend entsprechen, denn beide Arten haben 3 Typen von Zentralnadeln. Bei *C. faenofurca* umgeben insgesamt 12 Zentralnadeln ein Trichterrohr. RISER (l. c.) nennt für *C. erotica* jedoch keine Zahl der Zentralnadeln und erwähnt auch nicht die Existenz eines Trichterrohres, ein solches ist aber in einer Abbildung bei RISER (1981, fig. 1 K) zu sehen.

Zusammenfassung

Coelogynopora faenofurca nov. spec. wird von der deutschen Nordseeküste beschrieben. *C. erotica* von der amerikanischen Atlantikküste ist die nächstverwandte Art der neuen Species.

C. faenofurca wurde bisher ausschließlich in den Wohnröhren von *Arenicola marina* gefunden.

Abkürzungen in den Abbildungen

bs	Bursa	ph	Pharynx	te	Hodenfollikel
c	Gehirn	st	Statocyste	vg	Vesicula granulorum
ge	Germar	sti	Stilettnadeln	vi	Vitellarien
kd	Kittdrüsen	stl	Statolith	vs	Vesicula seminalis
ksd	Kornsekretdrüsen				

Literatur

REISE, K. (1984): Free-living Plathelminthes (Turbellaria) of a marine sand flat: an ecological study. Microfauna Marina **1**: 1–62.

REISE, K. & P. AX (1979): A meiofaunal "Thiobios" limited to the anaerobic sulfid system of marine sand does not exist. Mar. Biol. **54**: 225–237.

RISER, N. W. (1981): New England Coelogynoporidae. Hydrobiologia **84**: 139–145.

SCHERER, B. (1985): Annual dynamics of a meiofauna community from the "sulfide layer" of a North Sea sand flat. Microfauna Marina **2**: 117–161.

SOPOTT-EHLERS, B. (1976): Interstitielle Macrostomida und Proseriata (Turbellaria) von der französischen Atlantikküste. Mikrofauna Meeresboden **60**: 1–35.

— (1985): Zwei neue Proseriata (Plathelminthes) von der französischen Atlantikküste. Microfauna Marina **2**: 389–396.

Dr. Beate Sopott-Ehlers
II. Zoologisches Institut und Museum der Universität Göttingen
Berliner Straße 28, D-3400 Göttingen

Ultrastructure of the photoreceptors in certain larvae of the Annelida

Thomas Bartolomaeus

Contents

Abstract	191
A. Introduction	192
B. Material and Methods	192
C. Results	193
Spirorbis spirorbis	195
Autolytus prolifer	198
Pectinaria koreni	205
Lanice conchilega	205
D. Discussion	209
Zusammenfassung	213
Literature	213

Abstract

The eye-spots of the trochophore of *Spirorbis spirorbis*, of the newly hatched larva of *Autolytus prolifer* and of the tubicolous planctonic larvae of *Pectinaria koreni* and *Lanice conchilega* are investigated ultrastructurally. The ocelli are composed of two kinds of cells, i. e. cup-shaped pigment cells and sensory cells; one of each are present in *Lanice conchilega* and *Autolytus prolifer* and two of each in *Spirorbis spirorbis*. In *Pectinaria koreni* the eye is composed of one pigment cell and two differing sensory cells. Generally the sensory cells are rhabdomeric, but in *Autolytus prolifer* a rudimentary cilium is situated between the rhabdomeric microvilli and in *Pectinaria koreni* tubules originating from the endoplasmic reticulum extend into the microvilli of one of both sensory cells. In *Autolytus prolifer* the pigment cell secretes an extracellular lense. A unique intracellular accumulation of mitochondria directed towards the incident light is found in each sensory cell of *Spirorbis spirorbis*. One pair of photoreceptors, each consisting of one pigment cell and one rhabdomeric sensory cell with a rudimentary cilium are assumed for the larvae in the ground pattern of annelids. Lenses and higher numbers of cells are apomorphic characteristics.

A. Introduction

Photoreceptors of adult annelids have been investigated quite intensely (s. EAKIN & HERMANS 1988), but only a relatively small number of investigations deal with larval photoreceptors (s. table 1). An ultrastructural analysis of the paired photoreceptors of the trochophore of *Anaitides mucosa* (Phyllodocidae) led to considerations on the eyes of the trochophore in the groundpattern of annelids (BARTOLOMAEUS 1987). It is hypothesized that the trochophoral photoreceptor in the last common stem species of annelids consisted of one cup-shaped pigment cell (supporting cell) and one monociliated rhabdomeric sensory cell. This paper deals with ultrastructural data on the eyes in the larvae of *Spirorbis spirorbis* (Serpulidae), *Autolytus prolifer* (Syllidae), *Pectinaria koreni* (Terebellida) and *Lanice conchilega* (Terebellida) which support the above-mentioned hypothesis.

This investigation was financially supported by the Akademie der Wissenschaften und der Literatur, Mainz. I would like to thank Dr. T. Kunert, who kindly corrrected the manuscript, and Mrs. K. Lotz for technical assistance.

B. Material and Methods

Adult *Spirorbis spirorbis* Linne, 1758, which live in calcareous tubes on *Fucus serratus*, breed their spawn inside the tube. In Summer 1989, *Spirorbis spirorbis* was collected on the isle of Helgoland and removed from the tubes to obtain the spawn, which then was reared in Petri-dishes. The hatching larvae were fixed for electron microscopy. Aulophora larvae of *Lanice conchilega* Pallas, 1776 as well as trochophores and older tubicolous larvae of *Pectinaria koreni* Malmgren, 1865 were collected from plancton samples, which were taken off the isle of Helgoland in summer 1989 and fixed for electron microscopy. Female pelagic *Autolytus prolifer* O. F. Müller, 1776 (Sacconereis) carrying their developing eggs ventrally were collected from the same plancton samples and kept alive until hatching of the larvae. These were directly fixed for electron microscopy and five days after hatching.

The larvae were fixed in 2,5 % glutaraldehyde buffered in 0.1 M sodium cacodylate (pH 7.2) for 60–90 minutes at 4° C, rinsed in the same buffer and postfixed in 1 % OsO_4 buffered in 0.1 M sodium cacodylate at 4° C for

60 minutes. The larvae then were dehydrated in an acetone series, embedded in araldite and cut into a complete series of silver interference sections (65–75 nm) with a diamond knife on a REICHERT Ultracut microtome. The sections were kept on formvar-covered single slot copper grids, automatically stained with uranyle acetate and lead citrate in an LKB Ultrostainer and examined in ZEISS EM 10 B and EM 900 electron microscopes. Two larvae of each species have been investigated. The position and the ultrastructure of the eyes were reconstructed according to complete serial sections.

C. Results

The photoreceptors investigated here consist of two types of cells, i. e. cells which receive and percept optical information and cells which are a barrier to optical information. In this paper, the latter cells which contain pigment granules and which sometimes are described as supporting cells will be termed pigment cells. The perception of the optical information occurs in special sensory cells, which seem to never contain pigment and always exhibit a conspicious polarity, because the photosensitive area with microvilli and/or cilia is opposite to the perikaryon and the neurite. Mikrovilli and cilia extend into an invagination of the cup-shaped pigment cell, the optical cavity.

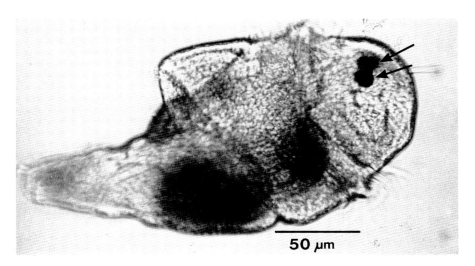

Fig. 1: Larva of *Spirorbis spirorbis* with one pair of eyes (arrows).

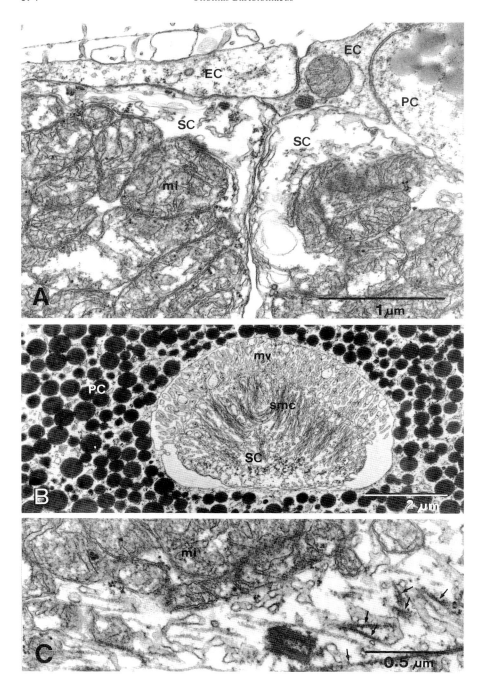

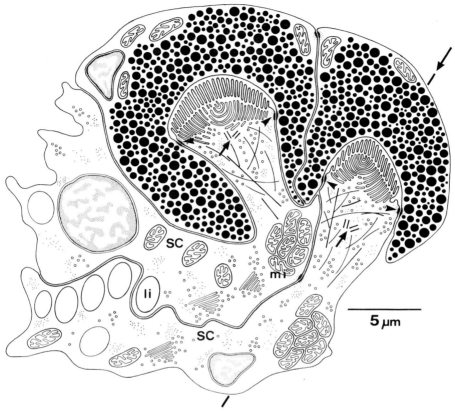

Fig. 3: Larva of *Spirorbis spirorbis*. Cross section of the eye redrawn from a series of EM micrographs. The eye consists of two sensory cells (SC) and two pigment cells. Arrows point at the diplosome, arrowheads at belt desmosomes. The line indicates the direction of the longitudinal section in Fig. 5. li = lipid vesicles, mi = mitochondria.

Spirorbis spirorbis

When the trochophores of *Spirorbis spirorbis* hatch from the spawn two setigerous segments have already been developed. At this stage one pair of dorsally situated red pigmented eye spots can clearly be seen at the anterior end (Fig. 1). The eyes are situated directly underneath a flattened epidermal cell

◄ Fig. 2: Larva of *Spirorbis spirorbis*. A. The sensory cells (SC), which have an apical concentration of mitochondria (mi) are situated directly underneath epidermal cells (EC). B. Cross section of the sensory cell inside the pigmented cup. Submicrovillar cisternae (smc) are underneath the rhabdomeric microvilli (mv). C. Rootlet-like structures (arrows) are associated with the centriole of the diplosome. PC = pigment cell.

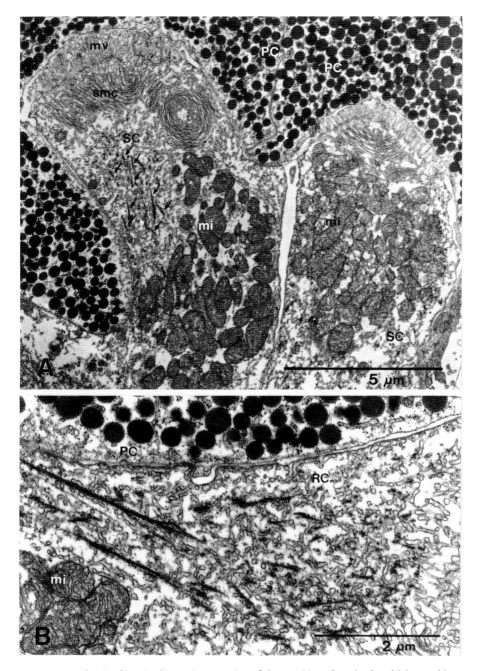

Fig. 4: Larva of *Spirorbis spirorbis*. A. Cross section of the eye. Note the mitochondrial assemblage (mi) inside the sensory cells (SC). Underneath the submicrovillar cisternae (smc) several rootlet-like fibers can be seen (arrows). B. Higher magnification of the presumed rootlets. mv = microvilli, PC = pigment cell.

(Fig. 2 A); there is no ECM (extracellular matrix) between these cells and the eyes. Each ocellus is composed of two pigment cells and two sensory cells (Fig. 3, 4 A). The eyes are inverted; i. e. the incident light has to pass the cell body before it is percepted.

Each pigment cell has the shape of an asymmetric cup with an inner diameter of max. 6 μm and an inner length ranging between 6 and 18 μm due to the asymmetrical shape. The nucleus and nearly all mitochondria are situated peripherally; the cell contains pigment granules of almost the same size (0.43 μm in

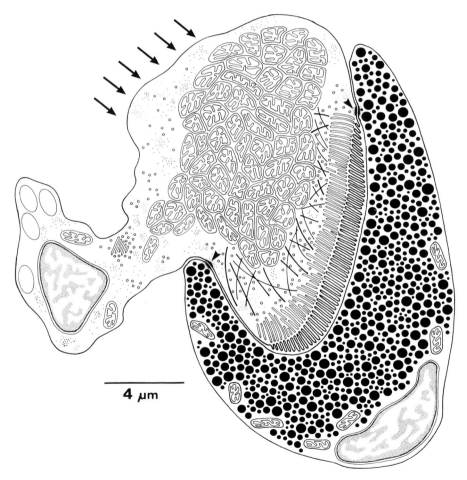

Fig. 5: Larva of *Spirorbis spirorbis*. Reconstruction of a longitudinal section of the inverse eye at the level indicated in Fig. 3. Arrow heads point at the belt desmosomes. The incident light (arrows) has to pass the mitochondrial assemblage before it is percepted.

diameter, n = 48, s = 0.04). Both pigment cells laterally border on one another and are connected by desmosomes.

Distally the sensory cells are next to one another and are connected by spot desmosomes. The perikarya are always situated outside the cup, which is formed by the pigment cells. The perikaryal cytoplasm contains a regularly shaped nucleus, large lipid vesicles, mitochondria, dictyosomes and numerous small vesicles. A relatively small cytoplasmic bridge (3–4 µm in diameter) connects the perikaryon with the main part of the sensory cell, which bears the presumably photosensitive rhabdomeric microvilli (Fig. 3, 5). Here, each of the sensory cells extends into its own optical cavity, i. e. the photosensitive parts of both sensory cells are separated from one another (Fig. 2 B, 4 A). The rhabdome consisting of microvilli which are 1 µm in length and about 70 nm in diameter, does not completely fill up the whole optical cavity. It solely extends along a relatively small area of about 5×10 µm^2 of the inner surface of the pigment cell (Fig. 2 B, 4 A). Underneath the rhabdomeric microvilli a layer of submicrovillar cisternae can be seen, which is followed by a netlike arrangement of several less distinctly striated, small electron dense fibers, which resemble ciliary rootlets. Here, the cytoplasm of the sensory cell is intensely vesiculated and contains a diplosome, which seems to be associated to the rootlet-like fibers (Fig. 2 C, 4 B). The angle between both centrioles is 35°.

This region is followed by a remarkable assemblage of far more than 100 tightly packed mitochondria. The distance between the mitochondria ranges between 12 nm in the center and 0.5 µm in the periphery of the assemblage. This mass of mitochondria is situated above the rhabdome and extends to the subepidermal surface of the sensory cell (Fig. 2 C, 4 B, 5). It thus completely covers the region of photoperception so that the incident light has to pass the mitochondrial assemblage before it reaches the rhabdomeric microvilli. It seems likely that the mitochondria modify the visual information and thus have a physical function in light perception. The number of mitochondria is much higher than in other cells of the animal.

Autolytus prolifer

The five day old larvae have four red pigmented eye spots, two lie dorsally and two laterally. During development, the lateral eyes appear a bit earlier than the dorsal ones. The eyes are situated at about the same level as the frontal part of the pharynx (Fig. 6 C). The eyes are situated directly underneath flat epidermal cells (Fig. 7 C). There is no ECM (extracellular matrix) between the eye and the overlying epidermal cells. Each photoreceptor is composed of one pigment cell and one sensory cell and possesses a lens (Fig. 8 A).

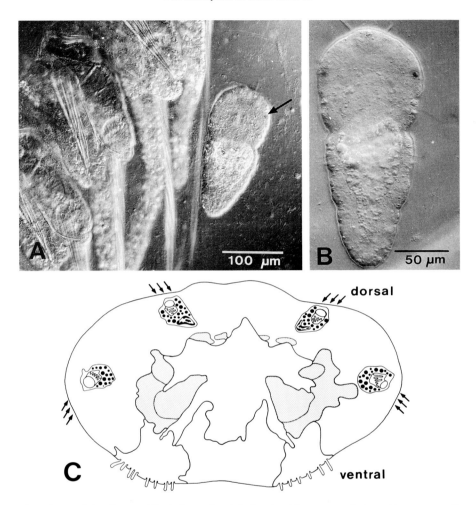

Fig. 6: Larva of *Autolytus prolifer*. A. Larva shortly before hatching, eyes are indicated by an arrow. B. Larva, shortly after hatching. C. Cross section of the larva. Nerve tissue is stippled, arrows indicate the direction of incident light.

The pigment cell of each photoreceptor is cup shaped and forms the lens (Fig. 7 B). Several short microvilli seem to enlarge that part of the cell surface which secretes the lens into the optical cavity, which is bordered by the pigment cell and the rhabdome of the sensory cell. The lens consists of amorphous electron-dense material. The lenses have a slightly oval shape measuring about 1.9 µm in diameter in the lateral eyes. Due to the delayed development the lenses

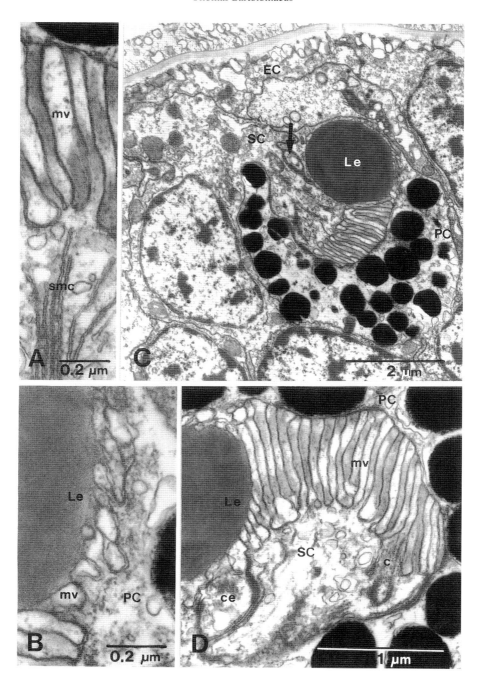

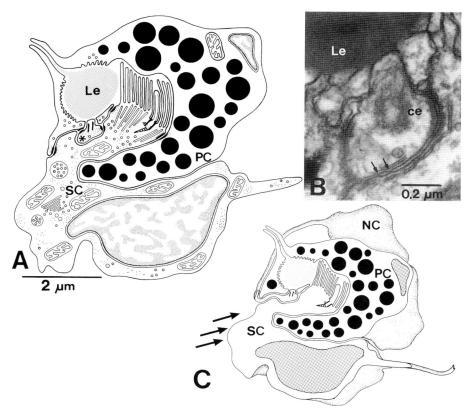

Fig. 8: Larva of *Autolytus prolifer*. A. Longitudinal section of the photoreceptor, redrawn from a series of EM-micrographs. Asterisk marks the neurite. B. Terminal region of the neurite, arrows point at tight juctions. C. Semi-schematic reconstruction of the eye and the perikaryon (NC) of the neurite. Arrows indicate the direction of the incident light. SC = sensory cell, PC = pigment cell, Le = lense, ce = centriole.

in the dorsal eyes each measure only 1.3 μm. A small, slender and hollow process of the cell proximally leads into the epidermis and connects the optical cavity with the cuticle. Fibrillar material can be seen inside this process indicating that the lens might be composed of the same material like the cuticle (Fig. 9 B–E). However, this process is a remnant of the original intraepidermal position of the

---

◄ Fig. 7: Larva of *Autolytus prolifer*, five days after hatching. A. Sensory cell. Rhabdomeric microvilli (mv) and submicrovillar cisternae (smc). B. Microvilli (mv) of the pigment cell (PC) seem to be involved in secreting the lens (Le). C. The eye is situated underneath epidermal cells (EC) and consists of one sensory cell (SC), one pigment cell (PC) and a lens (Le). Arrow points at a neurite. D. The sensory cell has a rudimentary cilium (c), the neurite contains a centriole (ce) terminally.

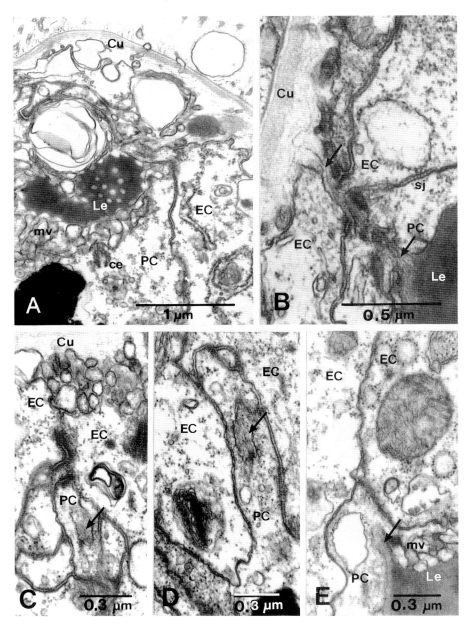

Fig. 9: Larva of *Autolytus prolifer*. A. Newly hatched larva. The pigment cell (PC) lies between epidermal cells (EC) and begins to secrete a lens (Le). B.–E. Five days after hatching. B. Dorsal eye. Epidermal cells cover the pigment cell, material of the lens (arrows) extends into the cuticle (Cu). C.–E. Ventral eye. Series of sections of a slender hollow protrusion of the pigment cell which runs apically contains fibrillar material (arrows) originating from the periphery of the lense. Distances C.–D. about 800 nm and D.–E. about 450 nm. mv = microvilli, sj = septate junctions.

pigment cell, which is described below (Fig. 9 A). The pigment cell contains pigment granules measuring 0.75 μm in diameter (n = 16, s = 0.1). The nucleus and the mitochondria are situated peripherally, just below the cell surface. A single centriole is located underneath the cell surface close to the lens-secreting microvilli (Fig. 9 A).

The perikaryon of the sensory cell is situated laterally to the opening of the pigmented cup. A small process (about 1 μm in diameter) connects the perikaryon to the presumably photosensitive part of the sensory cell. This area consists of microvilli, which are 0.9 μm in length and 70–80 nm in diameter. The longitudinal axes of the microvilli are oriented almost rectangularly to the incident light (Fig. 8 C). Between the microvilli, a short rudimentary cilium with a $9 \times 2 + 0$ axoneme is attached to the cell body by a basal body, which has a short striated rootlet and a lateral basal foot (Fig. 7 D). No accessory centriole is associated with the basal body. Underneath the cilium and the microvilli a layer of submicrovillar cisternae can be seen, which are presumed to be a derivative of smooth endoplasmic reticulum (Fig. 7 A). Mitochondria, small vesicles and endosomes are irregularly scattered in the cytoplasm of the sensory cell. Sensory cell and pigment cell are connected by belt desmosomes in the region of the rhabdome (Fig. 7 C, 8 A).

A small process of a nerve cell running between sensory cell and pigment cell contacts the lens to form a knop-like thickening with a centriole inside (Fig. 8 B). Beside this, the cytoplasm contains numerous small vesicles. The process is desmosomally connected to both photoreceptor cells. The cytoplasm of this process contains small vesicles. The perikaryon of this cell is situated dorso-caudally to the eye and is connected to several nerve cells of the cerebral ganglia (Fig. 8 C). Considerations on this nerve cell are discussed below.

In newly hatched larva of *Autolytus prolifer* the pigment cell is still situated between epidermal cells. The pigment cell widely opens up to the body surface. Several microvilli secrete irregularly arranged electron dense material, the prospective lense. The sensory cell is situated almost in the definite position. Obviously the pigment cell sinks into deeper cell layers during the development of the larva. As a remnant of the original wide opening towards the body surface, a small hollow protrusion containing fibrillar material remains and connects the pigment cell with the epidermal cell layer. In those stages fixed five days after hatching this protrusion is more prominent in the dorsal eyes, which appear later than the lateral ones during development (Fig. 9).

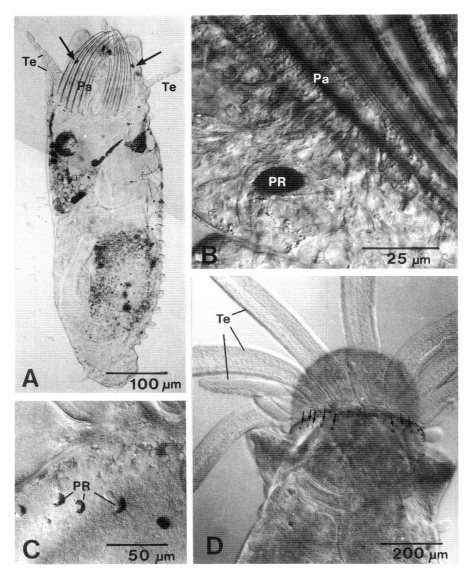

Fig. 10: A., B. Larva of *Pectinaria koreni*. A. Total view of the animal, arrows indicate the eyes. B. The eyes are situated laterally to the bases of the digging setae (Pa). C., D. Aulophora larva of *Lanice conchilega*. C. Several eyes (PR) at the basis of the tentacle crown. D. Head. Arrows point at the eyes. Te = tentacle.

Pectinaria koreni

Two different larval stages have been examined, i. e. the young trochophore, which has no chaetae and an older pelagic stage, which already lives in a hyaline tube and possess frontal digging chaetae (paleen), tentacles and parapodia. At both stages two black eye spots are discernable, which I could also observe in very young bottom stages (Fig. 10 A, B). Frontally, the eyes are located on the ventral side underneath the paleen and consist of one pigment cell and two sensory cells. The eyes are inverted.

The pigment cell is cup-shaped. The pigment granules differ in size reaching a maximal diameter of 1.5 µm in diameter. The margins of the cell do not contain any pigment granules. They are lobed and interdigitate with each other and the sensory cells, thus resembling a hollow sphere with a few small openings by which the photosensitive regions of the sensory cells enter the optical cavity. Therefore, always several cuts of the pigment cell can be seen in a series of longitudinal sections of the photoreceptor. The nucleus and the mitochondria are situated peripherally and the marginal cytoplasm contains a diplosome (Fig. 11 A, 12).

The perikarya of the sensory cell are situated right in front of the optical cavity. Both sensory cells are connected by spot desmosomes. The perikaryal cytoplasm contains mitochondria, vesicles, multivesiculated or multilamellate bodies (Fig. 12). The presumable photosentive region differs in both cells. One bears short, branched microvilli measuring about 1.5 µm in length and 170 nm in diameter. One or two membranous tubes, which are continous with the submicrovillar cisternae extend into these microvilli. The photosensitive region of the second sensory cell bears small, unbranched microvilli, which are about 4.8 µm in length and up to 90 nm in diameter and contain bundles of filaments. There is no layer of submicrovillar cisternae underneath the microvilli (Fig. 11 B). Each sensory cell has a diplosome which is situated close to the microvilli.

Lanice conchilega

A short note on the structure of the paired larval ocelli has been given by HEIMLER (1983). These eyes persist during further development and are identical to those of the planctonic aulophora larvae, which live in a hyaline tube. The larvae have several red eye-spots located at the bases of the tentacles (Fig. 10 C, D). The number of ocelli increases during development like the number on tentacles. Each ocellus consists of one cup shaped pigment cell and one sensory cell. The eyes are inverted (Fig. 14 C).

The pigment cell is nearly spherical with a optical cavity measuring about

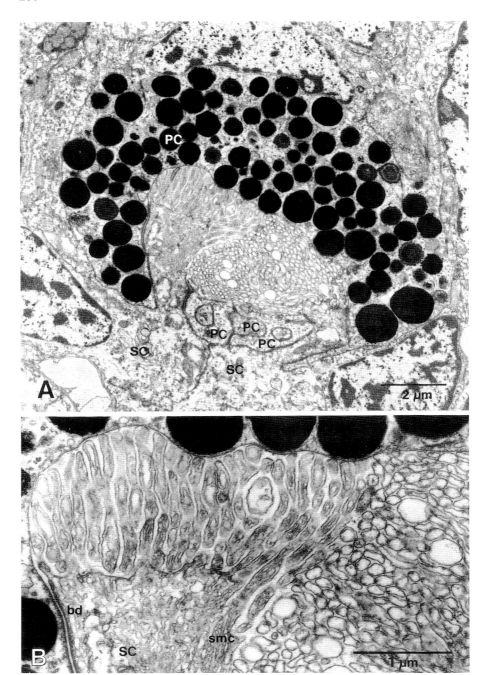

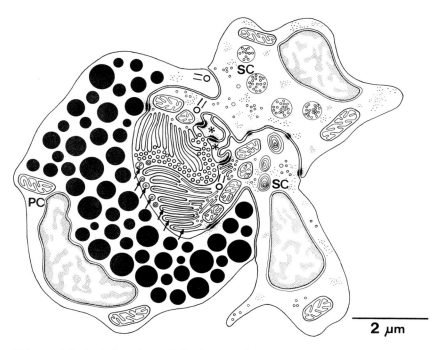

Fig. 12: Larva of *Pectinaria koreni*. Longitudinal section of the eye, redrawn from a series of sections. Note the tubules inside the microvilli in one of the two sensory cells (arrows). Asterisks mark the interdigitation of the pigment cell (PC). SC = sensory cell.

4 μm in diameter. The margins of the pigment cell are lobed and interdigitate to form a lateral seam (Fig. 13 A, 14). With the exception of this marginal area the cytoplasm of the cell contains pigment granules ranging between 0.75 μm and 1.1 μm in diameter. The nucleus and the mitochondria are situated peripherally, a diplosome can be seen underneath the inner cell membrane.

The perikaryon of the sensory cell is situated slightly laterally to the optical cavity. The perikaryal cytoplasm contains vesicles, a few small vacuoles and a few mitochondria and appears relatively electron-bright. Close to the nucleus a few endosomes can be found. The presumed photosensitive region extends into the spherical optical cavity and consists of a tuft of rhabdomeric microvilli, which are 1.9 μm in length and about 75 nm in diameter. At the bases of the

◄ Fig. 11: A., B. Larva of *Pectinaria koreni*. A. Longitudinal section of the eye. Asterisk mark the interdigitating cytoplasmic areas of the pigment cell (PC). B. Tubules of the submicrovillar cisternae (smc) extend into the microvilli of one sensory cell. bd = belt desmosomes, e = endosome, SC = sensory cell.

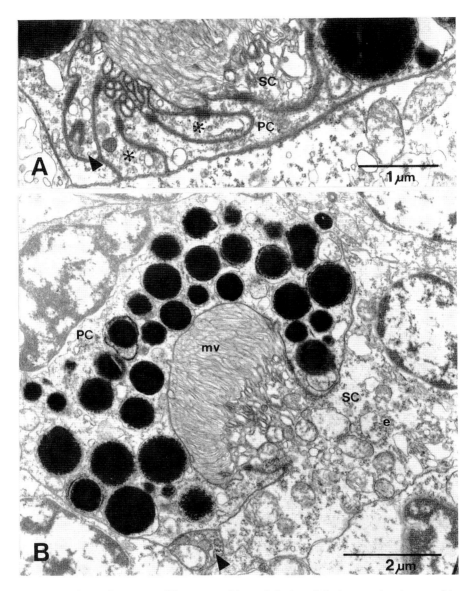

Fig. 13: Aulophora of *Lanice conchilega*. A. Asterisks mark the interdigitating cytoplasmic areas of the pigment cell (PC). A neurite (arrow head) extending into this cell contains a diplosome and bears microvillar structures. B. Longitudinal section of the eye. The neurite from A. (arrow head) runs parallelly to the pigment cell. e = endosome, SC = sensory cell.

microvilli the cytoplasm contains numerous vesicles and stacks of submicrovillar cisternae lying longitudinally to the main axis of the microvilli. Several mitochondria and a diplosome are situated close to this region. At about the level of the rhabdome belt desmosomes connect the sensory cell to the pigment cell (Fig. 13 B, 14).

A small process of a nerve cell, which is situated laterally and behind to the ocellus, runs parallelly to the pigment cell and pierces the cell in the region, where the cell margins interdigitate (Fig. 14 C). The process extends into the optical cavity and branches into short microvillar structures. A centriole with an accessory centriole is situated at the basis of these structures. Subapically belt desmosomes link this process to the pigment cell (Fig. 13 A). Numerous small vesicles and a few small mitochondria can be found within the cytoplasm of this process (Fig. 13 B).

A comparable nerve cell process has been found in the larva of *Autolytus prolifer*. Considerations on these nerve cell processes will be discussed below.

D. Discussion

Until now only a small number of eyes in the trochophore larvae have been investigated ultrastructurally. The results concerning the number of different cells and their structure are summarized in table 1.

According to the hypothesis of PIETSCH & WESTHEIDE (1985) a homology of the cerebral eyes is highly probable within the Eumetazoa. On the other hand the authors point at the basic difficulties in the application of the eye ultrastructure to phylogenetic implications. In the case of the eyes in annelid larvae a special difficulty arises from the fact that the photoreceptors in some larvae are but already functioning developmental stages of the adult ocelli, whereas in other larvae the eyes degenerate during development (EAKIN & WESTFALL 1964, HOLBOROW & LAVERACK 1972, VANFLETEREN 1982).

In two species, i. e. in *Autolytus prolifer* and in *Lanice conchilega*, a neurite is associated to the sensory cell. In *L. conchilega* the neurite contains a diplosome and bears short microvillar protrusions terminally. In *A. prolifer*, the neurite contains a centriole terminally which gives rise to a ciliary rudiment. Two hypotheses on the function of these nerve cells seem possible:
- The cells have an inhibitory function in order to modify the response of the sensory cell to optical information or
- the cell is beginning to change into an additional sensory cell.

Evidence for the persistence and enlargement of the larval eye would maintain the latter hypothesis. On the other hand this hypothesis presumes that the

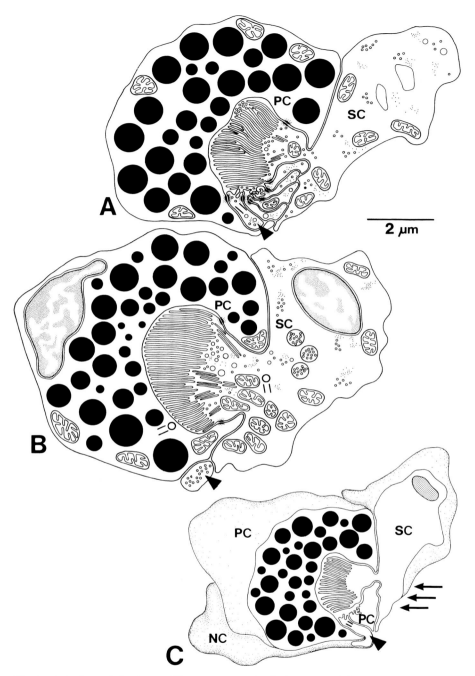

Fig. 14: Aulophora of *Lanice conchilega*. A., B. Longitudinal sections of the eye as redrawn from EM-micrographs. Arrow head marks the neurite from a nerve cell. (Distance A.–B. approx. 1,3 µm). C. Semi-schematic drawing of the eye and the nerve cell (NC). Arrows indicate the direction of the incident light. PC = pigment cell, SC = sensory cell.

Table 1: Photoreceptors in the larvae of Annelida

Taxon	Sensory cell (SC)	Pigmented cell (PC)	Author
Syllis amica	1–2 SC, rud. cilium, rhabdomeric	1 PC, secr. a lense	VERGER-BOCQUET 1983
Odontosyllis ctenostoma	1–2 SC, rud. cilium, rhabdomeric	1 PC, secr. a lense	VERGER-BOCQUET 1983
Autolytus prolifer	1 SC, rud. cilium, rootlet, rhabdomeric	1 PC, secr. a lense, rud. cilium	this study
Anaitides mucosa	1 SC, rud. cilium, ac, rootlet, rhabdomeric	1 PC	BARTOLOMAEUS 1987
Harmothoe imbricata	1–2 SC, rhabdomeric	1 PC	HOLBOROW & LAVERACK 1972
Neanthes succinea	1 SC, centriole, rootlet, rhabdomeric	1 PC*	BAKIN & WESTFALL 1964
Polygordius cf. appendiculatus	1 SC, rud. cilium, rootlet, rhabdomeric	2 PC	BRANDENBURGER & EAKIN 1981
Lanice conchilega	1 SC, diplosome, rhabdomeric	1 PC, interdig. marg., diplosome	this study
Pectinaria koreni	2 SC, diplosome, rhabdomeric	1 PC, interdig. marg., diplosome	this study
Spirorbis spirorbis	2 SC, diplosome, rootlets, rhabdomeric	2 PC	this study
Serpula vermicularis	1 SC, centriole, rootlets, rhabdomeric	1 PC, short mv	MARSDEN & HSIEH 1987
Spirobranchus giganteus	1 SC, centriole, rootlets, rhabdomeric	1 PC, short mv	MARSDEN & HSIEH 1987, SMITH 1984

Secretion of an extracellular lens by the pigment cell occurs in the first three species which are representatives of the Syllidae. In *Neanthes succinea* the pigment cell contains in intracellular accumulation of lens-like granules (asterisk). Interdigitation margins of the pigment cell are found in the two terebellid species, *P. auricoma* and *L. conchilega*. ac = accessory centriole, mv = microvilli, rud. = rudimentary.

sensory cells in the photoreceptor are derivatives of the nervous system, i. e. nerve cells which grow towards the pigment cell and develop a photosensitive region. Actually, the development of the eye in *Autolytus prolifer* emphasizes this assumption. The pigment cell is intraepidermal and sinks into deeper cell layers during larval development. A small hollow extension leading to the cuticle reminds of the former intraepidermal position of the pigment cell. The sensory cell seems to originate from the developing cerebral ganglia and to induce the transition of the pigment cell.

Based on an investigation into the ultrastructure of the photoreceptor in the larva of *Anaitides mucosa* (BARTOLOMAEUS 1987), I hypothezised that one pair of pigmented cup ocelli with one pigment cell and one sensory cell belong to the larva in the ground pattern of annelids. The sensory cell is rhabdomeric and possesses a rudimentary cilium with a 9 × 2 + 0 axoneme, an accessory centriole and a rootlet. Structural features in the eyes of the larva, which differ from this plesiomorphic design and which are restricted to one or several groups within annelids must therefore represent apomorphic characteristics.

The extracellular lense, which has been found in the eyes of larval Syllidae, surely represents such a characteristic (Table 1). Until today such structure has not been described for any other annelid larva, although lenses occur in the adults of different annelid taxa (see EAKIN & HERMANS 1988 for an evolutionary evaluation). It therefore seems likely that the existence of a lens secreted by the pigment cell in the eye of the larva represents an apomorphy of the Syllidae or of a taxon within the Syllidae.

An assemblage of mitochondria which has to be passed by the incident light in the trochophore of *Spirorbis spirorbis* has never been described before in annelid larva. A comparable concentration of mitochondria can be seen in the tentacular ocelli of the sabellid *Branchiomma vesiculosum* (DRAGESCO-KERNEIS 1979), which additionally has a special lens-cell situated in front of the sensory cell. Whatever the exact function of such an mitochondrial assemblage may be, it seems obvious that it modifies the incident light.

Nearly all trochophore larva have only one cup-shaped pigment cell (table 1). This cell surrounds the photosensitive part of the sensory cell and exhibits vesicles of different sizes, which contain electron-dense material. This result corresponds to the hypothesis that eyes of the trochophore in the groundpattern of annelids have had only one pigment cell. A higher number of pigment cells in the larval eyes, like two cells in the larva of *Spirorbis spirorbis* therefore must be an apormorphic character. There is an remarkable interdigitation between the pigment cell and sensory cell in the eyes of both terebellid species. Additional information on the ultrastructure of the photoreceptors in larvae of representatives of this taxon is needed to decide whether this is an apormorphic characteristic.

The sensory cells of the investigated species are rhabdomeric. Only in the larva of *Autolytus prolifer* a rudimentary cilium with a short rootlet has been found, whereas an accessory centriole is lacking. In the other three species a diplosome can be seen underneath the apical photosensitive microvilli. Such a diplosome is also characteristic of the pigment cell as well as of other cells of these species. Because we do not know, how the centriole is effected to induce a ciliary axoneme and thus to become a basal body, it is principially impossible to decide, whether the diplosome hints at the former existence of a cilium in the sensory cell of an ancestor or not. On the other hand, in many species we find several ciliary rootlets associated to the diplosome (table 1). If a rhabdomeric sensory cell with a rudimentary cilium, an acessory centriole and a rootlet is presumed for the larvae in the ground pattern of annelids, the existence of these rootlets will easily be explained as plesiomorphic.

Zusammenfassung

Die Pigmentbecherocellen der Trochophora von *Spirorbis spirorbis*, der Larve von *Autolytus prolifer*, der planktonischen röhrenbewohnenden Larve von *Pectinaria koreni* und der Aulophora-Larve von *Lanice conchilega* werden ultrastrukturell untersucht. Die Ocellen bestehen aus zwei unterschiedlichen Zellformen, becherförmigen Pigmentzellen und Rezeptorzellen, die in diese hineinragen. Bei *Lanice conchilega* und bei *Autolytus prolifer* bestehen die Augen aus einer Rezeptor- und einer Pigmentzelle, bei *Spirorbis spirorbis* aus jeweils zwei Zellen und bei *Pectinaria koreni* aus einer Pigmentzelle und zwei unterschiedlichen Rezeptorzellen. Die Rezeptorzellen besitzen rhabdomerische Mikrovilli, lediglich bei *Autolytus prolifer* ist zusätzlich ein rudimentäres Cilium ausgebildet und bei *Pectinaria koreni* ragen Tubuli, die von submicrovillären Cisternen ausgehen, in die Mikrovilli einer der beiden Rezeptorzellen. Bei *Autolytus prolifer* ist darüber hinaus eine acelluläre Linse ausgebildet, die von der Pigmentzelle abgeschieden wird. Eine ungewöhnliche Ansammlung von Mitochondrien, die vom einfallenden Licht passiert werden müssen, befindet sich in jeder Rezeptorzelle von *Spirorbis spirorbis*. Für die Larve im Grundmuster der Anneliden wird ein Paar von Photorezeptoren angenommen, die jeweils aus einer Pigmentzelle und eine rhabdomerischen Sinneszelle mit einem rudimentären Cilium bestehen. Linsenbildungen und eine erhöhte Anzahl von Pigment- oder Sinneszellen stellen apomorphe Zustände dar.

Literature

BARTOLOMAEUS, T. (1987): Ultrastruktur des Photorezeptors der Trochophora von *Anaitides mucosa* Oersted (Phyllodocida, Annelida). Microfauna Marina **3**: 411–418.

BRANDENBURGER, J. L. & R. M. EAKIN (1981): Fine structure of ocelli in the larvae of an archiannelid, *Polygordius*, cf. *appendiculatus*. Zoomorphology **99**: 23–36.

DRAGESCO-KERNEIS, A. (1979): Sur la regeneration de la tache oculaire du papche de *Branchiomma vesiculosum* (Montagu), Annelide, Polychete. C. R. Acad. Sci. (Paris) **288 D**: 1179–1182.

EAKIN, R. M. & C. O. HERMANS (1988): Eyes. In: The ultrastructure of the Polychaeta (eds.: W. WESTHEIDE & C. O. HERMANS). Microfauna Marina **4**: 135–156.

EAKIN, R. M. & J. A. WESTFALL (1964): Further observations on the fine structure of some invertebrate eyes. Z. Zellforsch. **62**: 310–332.

HEIMLER, W. (1983): Untersuchungen zur Larvalentwicklung von *Lanice conchilega* (Pallas) 1776 (Polychaeta, Terebellomorpha). Teil III: Bau und Struktur der Aulophora-Larve. Zool. Jb. Anat. **110**: 411–478.

HOLBOROW, P. L. & M. S. LAVERACK (1972): Presumptive photoreceptor structures of the trochophore of *Harmothoe imbricata* (Polychaeta). Mar. Behav. Physiol. **1**: 139–156.

MARSDEN, J. R. & HSIEH, J. (1987): Ultrastructure of the eyespot in three polychaete trochophore larvae (Annelida). Zoomorphology **106**: 361–368.

PIETSCH, A. & W. WESTHEIDE (1985): Ultrastructural investigations of presumed photoreceptors as a means of discrimination and identification of closely related species of the genus *Microphthalmus* (Polychaeta, Hesionidae). Zoomorphology **105**: 265–276.

SMITH, R. S. (1984): Novel organelle associations in photoreceptors of a serpulid polychaete worm. Tissue & Cell **16**: 951–956.

VANFLETEREN, J. R. (1982): A monophyletic line of evolution? Ciliary induced photoreceptor membranes. In: Visual cells in evolution (ed: J. A. WESTFALL). Raven Press, New York: 107–136.

VERGER-BOCQUET, M. (1983): Etude infrastructurale des organes photorécepteurs chez les larves de deux Syllidiens (Annelides, Polychetes). J. ultrastr. Res. **84**: 67–72.

Dr. Thomas Bartolomaeus
II. Zoologisches Institut und Museum der Universität Göttingen
Berliner Straße 28, D-3400 Göttingen

Ultrastructure of the photoreceptor in the larvae of *Lepidochiton cinereus* (Mollusca, Plyplacophora) and *Lacuna divaricata* (Mollusca, Gastropoda)

Thomas Bartolomaeus

Contents

Abstract	215
A. Introduction	216
B. Material and Methods	216
C. Results	216
Lepidochiton cinereus	216
Lacuna divaricata	225
D. Discussion	229
Zusammenfassung	234
Literature	235

Abstract

The pelagic larva of *Lepidochiton cinereus* has one pair of dorsally situated intraepidermal eyes. These everse ocelli are each composed of four sensory cells and six pigment cells. Both kinds of cells are slender and have two or three cilia and numerous microvilli. During the formation of the shell plates the eyes shift into a ventral position. Thus, postlarval bottom stages, which already have 8 shell plates, possess one pair of ventro-laterally situated eye cups with sensory and pigment cells extremely interdigitating. Apically both bear long microvilli and cilia, which are tightly packed in the centre of the eye-spot. The veliger larva of *Lacuna divaricata* has one pair of subepidermal eye-spots. These ocelli consist of one pigment cell, one rhabdomeric and one ciliary sensory cell and one corneal cell which bears a lens. The rhabdomeric sensory cell has several long rhabdomeric microvilli and one cilium. The ciliary sensory cell bears several cilia and short irregularly shaped microvilli. A comparison within the Mollusca leads to the assumption that an acellular lens and ciliary sensory cells must have been evolved within gastropods. Lensless intraepidermal eye-spots consisting of several pigment and sensory cells are hypothesized to represent the photoreceptor of at least the larva in the ground pattern of the Eumollusca.

A. Introduction

Electron microscopical studies of photoreceptors within the Mollusca have mostly been done on adults of the Gastropoda and Cephalopoda (EAKIN & WESTFALL 1964, TONOSAKI 1965, 1967, EAKIN et al. 1967, HUGHES 1970 A, EAKIN et al. 1980) or on aesthetes of the Polyplacophora (BOYLE 1969, FISCHER 1978, 1979). There are only a few investigations on the ultrastructure of larval ocelli (HUGHES 1970 B, ROSEN et al. 1979, CROWTHER & BONAR 1980, FISCHER 1980, CHIA & KOSS 1983, GIBSON 1984, BUCHANAN 1986). This paper analyses the ultrastructure and development of the eye-spots of *Lepidochiton cinereus*, which have been described in a juvenile stage by FISCHER (1980). In addition, the ultrastructure of the eye-spots in *Lacuna divaricata* is investigated. A comparison of these results and data from other publications permit hypotheses on the photoreceptors in the larva of the ground pattern of the Mollusca.

This investigation was financially supported by the Akademie der Wissenschaften und der Literatur, Mainz. Dr. T. Kunert kindly corrected the manuscript.

B. Material and Methods

Spawn of *Lacuna divaricata* Fabricius, 1780, was collected on the isle of Helgoland in early spring 1991 from *Laminaria digitata* and kept in the laboratory of the BAH Helgoland until hatching of the veliger larvae, which were fixed about 6 hours later. For rearing of the larvae of *Lepidochiton cinereus* Linnaeus, 1767, and for electron microscopical preparations see BARTOLOMAEUS (1989). Complete series of sections from two individuals of each stage of *L. cinereus* and from one veliger larva of *L. divaricata* were examined in a ZEISS 10 B electron microscope.

C. Results
Lepidochiton cinereus

The red eye-spots of the larvae are situated laterally below the prototroch. Each eye consists of six pigment cells and four sensory cells. Like the surrounding epidermal cells, both, sensory and pigment cells, contact the outer medium

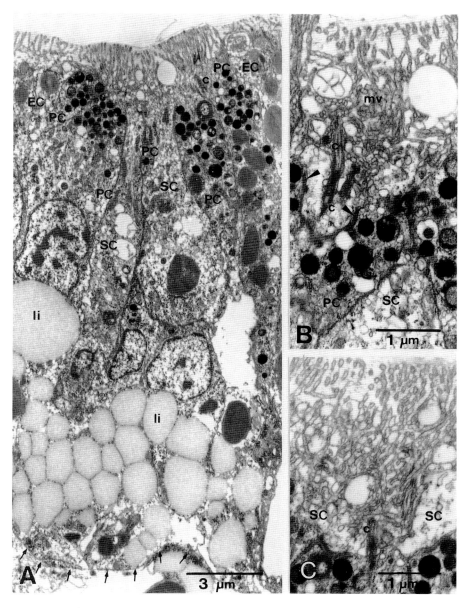

Fig. 1: *Lepidochiton cinereus.* Photoreceptor in the larva. A. The photoreceptor is situated between epidermal cells (EC) and rests on the subepidermal ECM (arrows). B. The sensory cells (SC) which bear cilia (c) and microvilli (mv) overlap the pigment cells (PC) to which they are connected by desmosomes (arrow-heads). C. The pigment cells are ciliated. li = lipid vesicle.

apically (Fig. 1 A). In juvenile bottom stages of *L. cinereus* the eyes are situated in the outer wall of the pallial rim. Epidermal cells completely overlap the eyes (Fig. 3 B, 6 C). The pigment cells have a cup-like arrangement. The sensory cells and pigment cells are compressed and interdigitate intensely (Fig. 5).

In the larva the pigment cells are small and slender measuring 19 µm in length and 2–3.5 µm in diameter. The pigment granules occupy the distal sixth of the cells and rather form a horizontal pigmented plate (up to 3 µm in height) than a pigmented cup. The regularly shaped nucleus is situated medially and mitochondria are scattered throughout the cytoplasm. In the basal half of the cells the cytoplasm is almost completely filled with huge lipid vesicles (Fig. 1 A). In the apical half of the cell the cytoplasm contains numerous small vesicles, ovoid and electron-grey vesicles, and vesicles containing flocculent material. The electron-dense pigment granules range between 0.3 and 0.45 µm in diameter and some seem to contain peripherally arranged osmiophilic material with a central core of electron-grey material (Fig. 1 B). The cell apex is lobed, numerous microvilli, which are 1.2 µm in length and about 95 nm in diameter extend from the surface into the surrounding medium (Fig. 1, 2). Filiform material can be seen between the tips of the microvilli forming a kind of apical network (Fig. 1 B, C). Each pigment cell bears two or three short cilia each with an $9 \times 2 + 2$ axoneme. A basal body connects the axoneme to a short rootlet. Belt desmosomes connect the pigment cells to sensory cells and epidermal cells (Fig. 1 C). Basally the pigment cells rest on the subepidermal ECM (Fig. 1 A).

The slender sensory cells in the larva are situated between pigment cells. The shape of these cells resembles a mushroom and the wide apical area partly overlaps the pigment cells. Cilia and microvilli extend from the apical surface into the surrounding medium. The $9 \times 2 + 2$ axoneme of the cilia originates from a basal body, which has one striated rootlet (Fig. 1 B). Like in pigment cells, these cilia measure about 2.5 µm and are thus much shorter than those of the surrounding epidermal cells, which measure up to 25 µm. The microvilli are straight and measure 1.2 µm in length and 95 nm in diameter. Subapically the cells are constricted, but soon widen towards the medially situated nucleus. Huge lipid vesicles almost completely fill up the proximal cytoplasm. Mitochondria seem to be associated to the rootlets of the cilia, but can also be found in any other region of the cell. Small vesicles are characteristic of these cells. It is important to notice that the microvilli of the sensory cells as well as those of the pigment cells are longer than those of the epidermal cell, which solely measure 0.7 µm in length (Fig. 1, 2).

The juvenile bottom stages possess ellipsoid eyes situated subepidermally, which are 15 µm in length and 8 µm in diameter. The arrangement of the pigment cells resembles a cup. Pigment granules ranging between 0.3 and 0.7 µm

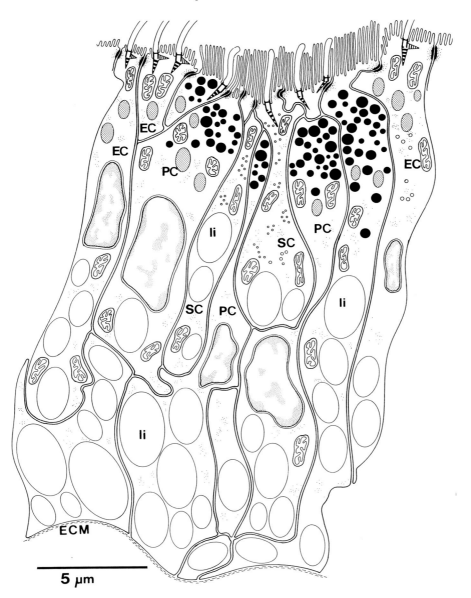

Fig. 2: *Lepidochiton cinereus*. Photoreceptor of the larva. Redrawn from several EM-micrographs. EC = epidermal cell, ECM = extracellular matrix, li = lipid vesicle, SC = sensory cell, PC = pigment cell.

in diameter occupy the apical regions of the pigment cells and form a pigmented band of up to 2 μm in thickness. The spherical optical cavity has a diameter of 4–5 μm and is completely filled up by microvilli and cilia, which originate from the pigment cells and the sensory cells (Fig. 3 B, 4, 5). The microvilli are much longer than in the eyes of the larvae; they measure about 4 μm in length and 95 nm in diameter. The cell apices of sensory and pigment cells are intensively folded and partly extend deeply into the cup. The cell bodies of both are lobed and interdigitate with neighbouring cells.

The sensory cells are dumbbell-shaped, i. e. the distal and the proximal areas of the cells are much broader than the central region, which is slender and neck-like. The wide hood-like apices of the sensory cells seem to form an incomplete inner layer partly overlapping the pigment cells, whereas the perikaryon is situated between the bases of the pigment cells (Fig. 3 A, 4). One sensory cell closes up the cup as its apex faces the central assemblage of microvilli and cilia. Amorphous, electron-bright material of about 120 nm in thickness can be seen to partly fill up a gap between the apex of this cell and the central microvilli. This material resembles a lens (Fig. 4 B, 5 B). The cytoplasm of both kinds of retina cells contains almost the same organelles as in the larvae. There solely seem to be more pigmented vesicles and less lipid vesicles in the pigment cells.

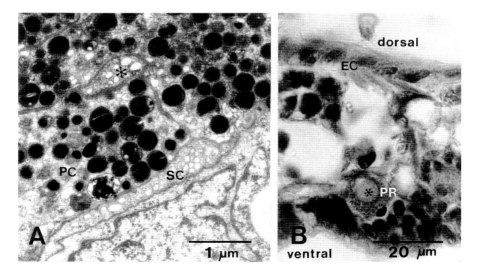

Fig. 3: *Lepidochiton cinereus*. Photoreceptor of a juvenile. A. Sagittal section. A slender process (asterisk) connects the photosensitive area to the perikaryon of the sensory cell (SC). B. Light-micrograph of the cross sectioned perinotum. Epidermal cells (EC) cover the eye (PR). Asterisk marks the central photosensitive area.

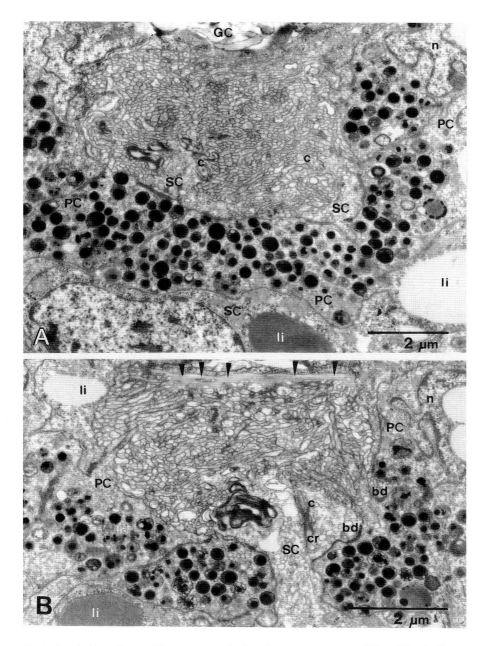

Fig. 4: *Lepidochiton cinereus*. Photoreceptor of a juvenile, cross sections. A. Microvilli and cilia (c) inside the optical cavity originate from the pigment (PC) and sensory cells (SC). B. A small process connects the photosensitive area to the perikaryon. Note the apical electron-grey mass (arrowheads). Distance A.–B. approx. 4.8 μm. bd = belt desmosome, cr = ciliary rootlet, GC = gland cell, li = lipid vesicle, n = nucleus.

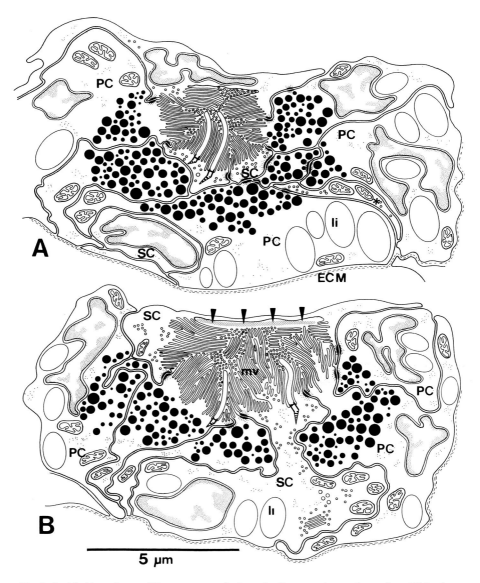

Fig. 5: *Lepidochiton cinereus*. Photoreceptor of a juvenile. Cross sections redrawn from EM-micrographs. A., B. The sensory cells (SC) partly overlap the pigment cells (PC). Arrow-heads point at apical amorphous material. ECM = extracellular matrix, li = lipid vesicle, mv = microvilli.

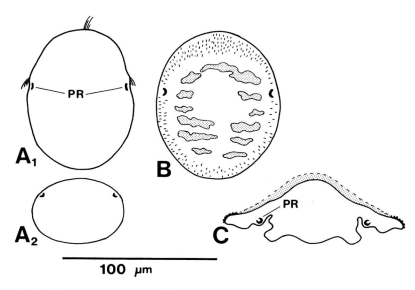

Fig. 6: *Lepidochiton cinereus*. Position of the photoreceptor (PR) in different developmental stages. A. In free swimming larva the eyes are situated dorsolaterally, underneath the prototroch (A_2 = cross section). B. During the formation of shell plates the eyes shift laterally and C. are finally situated ventro-laterally below the epidermis of the perinotum.

The nuclei of both are lobed and often situated peripherally. The retina cells are linked by belt desmosomes and rest on ECM (Fig. 4, 5).

Compared to the the photoreceptor of the larva, the eye of the juvenile bottom stage seems to be apico-basally compressed.

This drastic change in the morphology of the larval eye in *Lepidochiton cinereus* can be explained by the mode of development. The larvae of this species, which hatch from yolkrich eggs, only have a very short pelagic life lasting hardly longer than 60 minutes (Fig. 6 A). At the end of this period the larva sinks to the bottom and becomes a benthic juvenile with a ventral foot, a pallial rim and the dorsal shell anlage (Fig. 6 B). Thus, the larva, which has been almost round in cross section is dorso-ventrally compressed now. This morphological modification is mainly caused by the formation of dorsal shell plates. During shell formation epidermal material which has been in a lateral position shifts ventrally.

Accordingly the eyes, which have been situated laterally in the larva, become located in the outer wall of the pallial rim (Fig. 6 C). The slender sensory and pigment cells are compressed and epidermal cells overlap the eye. Later during development the eyes degenerate. At this stage their photoreceptive function has already been adopted by the shell eyes or aesthetes.

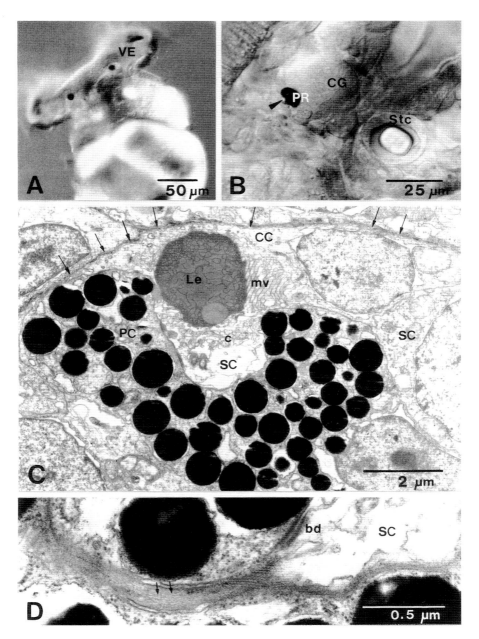

Fig. 7: *Lacuna divaricata*. Veliger-larva. A. Eyes are situated at the bases of the velum (VE), B. dorsally to the statocyst (Stc) closely to the cerebral ganglion (CG). Note the lens (arrow-head). C. The eye consists of one pigment cell (PC), two sensory cells (SC) and one corneal cell (CC). Arrows mark the ECM partly surrounding the eye. D. Neurite of the ciliary sensory cell. Arrows mark the neurofilaments. bd = belt desmosomes, c = cilium, Le = lens, mv = microvilli.

Lacuna divaricata

The veliger larva of *Lacuna divaricata* has one pair of black eyes situated at the basis of the velum dorsally to the mouth (Fig. 7 A). At the light microscopical level a small lens can be seen in front of the pigmented part of the eye (Fig. 7 B). The electron microscope reveals that the larval ocellus consists of four cells, i. e. one pigment cell, two sensory cells and one lens cell (Fig. 9). Each eye is situated underneath the epidermis and surrounded by ECM which also encloses the

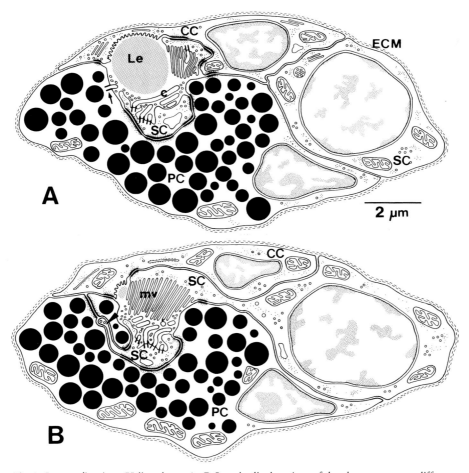

Fig. 8: *Lacuna divaricata*. Veliger-larva. A., B. Longitudinal sections of the photoreceptor at different levels as drawn from EM-micrographs. Distance A.–B. approx. 2 μm. There is one ciliary and one rhabdomeric sensory cell (SC). c = cilium, CC = corneal cell, ECM = extracellular matrix, Le = lens, mv = microvilli.

cerebral ganglion. The shape of the pigmented cell, the orientation of the sensory cells and the position of the eye allows the veliger larva of *L. divaricata* to perceive small sector of light only from an anterio-dorsal direction (Fig. 9).

The pigment cell exhibits a pericentrally situated cup-like invagination. The cell contains pigment granules measuring up to 0.95 µm in diameter. The long but relatively small nucleus and the mitochondria are situated peripherally (Fig. 8). A short rudimentary cilium without a discernable axoneme originates from the basal body and extends into the cup (Fig. 10 B). A small rootlet is attached to the basal body. There is no accessory centriole. Desmosomes connect the pigment cell to the remaining eye cells.

The perikarya of both sensory cells are situated laterally to the optical cavity. The nuclei of both are very large with a small amount of heterochromatin and are embedded in a thin layer of cytoplasm containing a few mitochondria, endosomes and vesicles (Fig. 7 C). One of both sensory cells narrows laterally to form a slender process with a knoplike thickening at the end. Here, the apical cell membrane bears numerous parallel microvilli, which measure 1.8 µm in length and 90 nm in diameter. They extend into the cuplike invagination of the pigment cell, the optical cavity. Between the microvilli a single short cilium with a 9 × 2 + 2 axoneme is attached to the cell by a basal body. There is no accessory

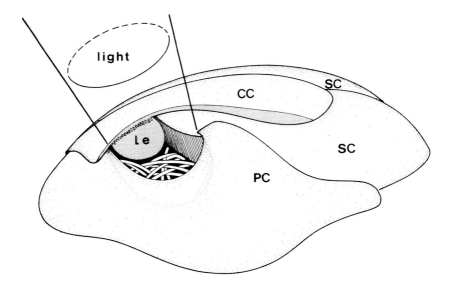

Fig. 9: *Lacuna divaricata*. Veliger-larva. Three dimensional scheme of the eye. The corneal cell (CC) and the lens (Le) are partly removed to demonstrate the organization of the optical cavity. As indicated, due to the cup shape of the pigmented cell (PC) only a small sector of light can be perceived. There are two different sorts of sensory cells (SC).

centriole. Belt desmosomes and septate junctions connect this presumed photosensitive part of the sensory cell to the lens cell and the pigment cell (Fig. 8, 11).

The perikaryon of the second sensory cell bears an extremely small neurite which pierces the pigment cell to form an irregularly shaped thickening at the bottom of the optical cavity (Fig. 11). Belt desmosomes and septate junctions connect the whole structure to the pigment cell. Several cilia with $9 \times 2 + 2$ axonemata emanate from this putative photosensitive part of the sensory cell. Each cilium has a basal body, no rootlet structures were found. The ciliary

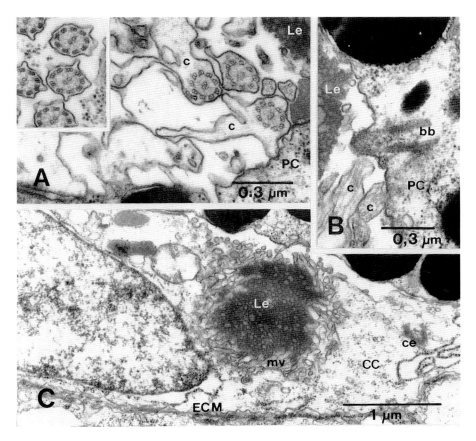

Fig. 10: *Lacuna divaricata*. Veliger-larva. A. Cilia (c) of the ciliary sensory cell have several irregular protuberances, the microtubular doublets have no dynein-arms. Inset: Cross sectioned cilia of the velum. Same magnification. B. The pigment cell (PC) has a rudimentary cilium. C. The lens (Le) is secreted by microvilli (mv). bb = basal body, CC = corneal cell, ce = centriole, ECM = extracellular matrix.

membrane forms lobes and does not surround the axoneme tightly. The peripheral microtubular doublets lack the dynein-arms which are characteristic of locomotory cilia (Fig. 10 A). These cilia must therefore be immotile. The microtubules are disorderly arranged towards the tip of the cilium (Fig. 11). Short

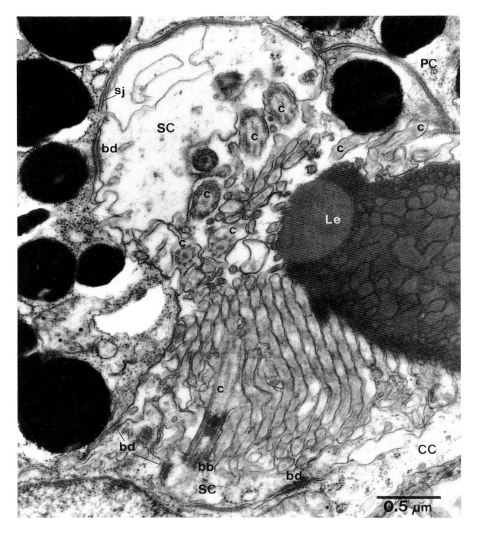

Fig. 11: *Lacuna divaricata*. Veliger-larva. Section of the optical cavity. There is one ciliated sensory cell and one rhabdomeric sensory cell, which also bears one cilium (c). Membranes can be seen inside the lens. bd = belt desmosomes, CC = corneal cell, Le = lens, PC = pigment cell, SC = sensory cell, sj = septate junction.

microvilli arise between the cilia and extend into the optical cavity. The cytoplasm of the thickening is relatively electron-bright and contains a large vacuole and a few small vesicles. The neurite leading to the perikaryon of the sensory cell is tightly packed with neurofibrils (Fig. 7 D).

Both sensory cells are desmosomally connected to nerve cells of the cerebral ganglia. At least one of the nerve cells bears a neurite which partly pierces the pigment cell to end blindly below the optical cavity. It is supposed that this cell may be a developing sensory cell.

The fourth cell in the eye of the veliger larvae of *Lacuna divaricata* is the corneal cell, which completely covers the optical cavity and partly overlaps the two receptor cells. With the exception of the laterally situated perinuclear region, the corneal cell is relatively flat (Fig. 8, 9). Microvilli, which are oriented towards the optical cavity secrete a lens of amorphous electron-grey material (Fig. 10 C, 11). Membranes of dissolving microvilli can be seen inside the lens. The lens is nearly spherical, measures about 3 µm in diameter and seems to extend deeply into the optical cavity. The cytoplasm of the corneal cell contains mitochondria, vesicles, several small vesicles with electron-dense contents resembling the lens material, and dictyosomes.

At the basis of the velum a few epidermal cells, which contain vesicles with electron-dense material inside can be found closely to the eye (Fig. 12). These cells secrete a cuticle. The material inside the vesicle is presumed to be identical to the pigment in the pigment cells of the eye.

D. Discussion

Eyes have been described in larvae of the Gastropoda (SMITH 1935, HUGHES 1970 b, CHIA & KOSS 1983, BUCHANAN 1986), the Bivalvia (see ROSEN et al. 1978) and the Polyplacophora (ROSEN et al. 1979, FISCHER 1980). In polyplacophores, the aesthetes substitute for the larval eyes during development, in some bivalves eyes in the epidermis of the mantle rim. In contrast to this, the eyes in the larvae of gastropods are retained through metamorphosis and develop into complex adult photoreceptors by increasing the number of sensory, pigment and corneals cells, but without any ultrastructural changes (FRETTER 1969, HUGHES 1970 b, GIBSON 1984). Due to this persistence the fine structure of gastropod eyes can be compared regardless of their developmental stage.

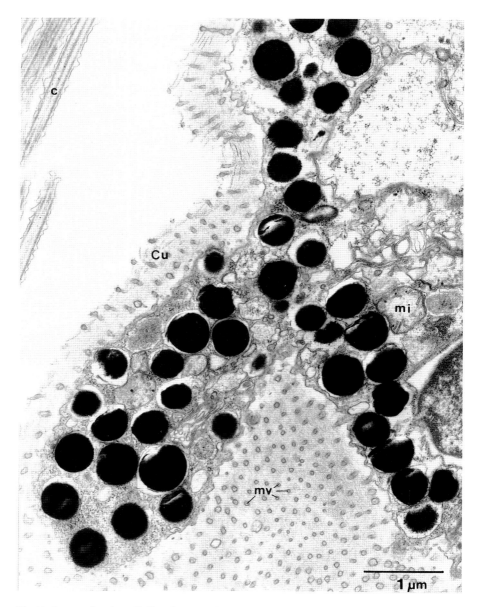

Fig. 12: *Lacuna divaricata*. Veliger-larva. Epidermal cell at the basis of the velum. The cell secretes a cuticle (cu) and contains electron-densely stained vesicles similar to the pigment cell. c = cilium, mi = mitochondria, mv = microvilli.

Gastropoda

Eyes possessing two different sorts of sensory cells, i. e. ciliated and rhabdomeric, like in the larval eyes of *Lacuna divaricata* have been described in the pulmonate *Helix aspersa* (EAKIN & BRANDENBURGER 1967) and in the opisthobranch *Aplysia punctata* (HUGHES 1970 a) as well as in *Onchidium* spec. (TONOSAKI 1967), *Ilyanassa obsoleta* (CROWTHER & BONAR 1980, GIBSON 1984) and *Littorina* spec. (HUGHES 1970 a). The Nudibranchia investigated so far have rhabdomeric sensory cells only (HUGHES 1970 a). It is supposed that ciliated sensory cells have been reduced in this taxon. This reduction is regarded as an apomorphy of the Nudibranchia. Accordingly, it is supposed that eyes with ciliary and rhabdomeric sensory cells represent the plesiomorphic condition within the Gastropoda.

Most of the pigment cells in gastropd eyes have microvilli and short rudimentary cilia extending from the cell apex into the optical cavity (HUGHES 1970 a, 1970 b, EAKIN et al. 1967, GIBSON 1984, CHIA & KOSS 1983). In *Haliotis discus* the pigment cells have extremely long cilia, each with a basal body and a rootlet (TONOSAKI 1967). It seems likely that pigment cells with microvilli, cilia and basal structures represent the plesiomorphic condition within gastropods.

An acellular lens has been observed in most gastropod eyes (RAVEN 1958). This lens is either secreted by the pigment and the corneal cells (EAKIN & BRANDENBURGER 1967, EAKIN et al. 1967, CROWTHER & BONAR 1980, GIBSON 1984) or solely by the corneal cell (HUGHES 1970 b, CHIA & KOSS 1983, this paper). The eye of *Haliotis discus* is an oval eye-cup with a pin-hole opening to the outer medium and without a lens (TONOSAKI 1967). The same has been observed in *Patella vulgata* (SMITH 1935). Since *Patella* and *Haliotis* are taxa with a large number of plesiomorphic characteristics within the Mollusca (HASZPRUNAR 1988), it is hypothesized that gastropod eyes primarily were lenseless and that a lens has been evolved within the Gastropoda. Moreover, the mode of lens development in gastropods differs fundamentally from that observed in the sister group of the Gastropoda, the Cephalopoda. Within cephalopods eyes with lenses are characterstic of representatives of the taxon Coleoidae including the fossil belemnoids and the Dibranchiata (BERTHOLD & ENGESER 1987). The lens of the dibranchiate cephalopod *Loligo pealii* is cellular. During development concentric cytoplasmic processes of the corneal cell aggregate "to form and 'onion bulb-like' structure which projects into the optic vesicle" (ARNOLD 1966, p. 535).

Corneal and pigment cells are supposed to be derivatives of the ectoderm (GIBSON 1984, BUCHANAN 1986). Presumably identical pigment granules in epidermis cells and pigment cells in *Lacuna divaricata* support this assumption.

In *L. divaricata*, it seems possible that nerve cells originating from the developing cerebral ganglia grow towards the pigment cells and differentiate into sensory cells. A neurite, which pierces the pigment cell and blindly ends below the optical cavity is interpreted as differentiating sensory cell.

For the veliger larvae in the ground pattern of gastropods cupshaped eyes consisting of several pigment and sensory cells are hypothesized. The eyes have no lenses, the pigment cells bear cilia and microvilli and the sensory cells have at least cilia and microvilli. It seems possible that distinct ciliary sensory cells have evolved from rhabdomeric sensory cells which had cilia within gastropods. Such cells represent the only sort of sensory cells in the eyes of larval polyplacophores and bivalves (see below).

Polyplacophora

The paired photoreceptors in the larvae of the Polyplacophora are intraepidermal and situated caudally to the prototroch. They consist of several pigment cells and one sensory cell in *Katharina tunicata* (ROSEN et al. 1979) or several sensory cells in *Lepidochiton cinereus*. In each species, the sensory cells have rhabdomeric microvilli and one or two cilia. In older, juvenile bottom-stages of *L. cinereus*, the photoreceptors have been shifted into a ventrolateral position and are situated in the perinotum and covered by epidermal cells (Fig. 3 B, FISCHER 1980). This shifting is presumably caused by the formation of shell plates after the short larval phase. The larval photoreceptors persist until late juvenile stages. Diverging data on the position of the photoreceptors in earlier papers (intraepidermal versus subepidermal, cit. from MOOR 1983) might merely be due to the investigations of different developmental stages.

Bivalvia

The larval eyes in some representatives of the Bivalvia are the first eyes that appear during embryonic development like in the larvae of polyplacophores. These eyes are intraepidermal, cupshaped and situated caudally to the prototroch. They are composed of pigment and sensory cells which contact the outer medium apically (RAVEN 1958). In *Mytilus edulis* the sensory cells are dumbbell-shaped, ciliated and partly overlap the pigment cells apically. The sensory cells are rhabdomeric and have one cilium, the pigment cells bear numerous microvilli (ROSEN et al. 1978).

Evolutionary evaluation

The following evolutionary considerations on mollusc photoreceptors are based on the diagram of phylogenetic relationships within the Mollusca which was substantiated by LAUTERBACH (1983), (Fig. 13).

The eyes in the larvae of gastropods, bivalves and polyplacophores are presumed to be homologous although they are innervated differently (ROSEN et al. 1979). Those of the larvae of bivalves and gastropods are connected with the cerebral ganglion whereas those of the polyplacophore larvae are connected with the lateral nerve cords. Due to the proposed homology, paired cup-shaped intraepidermal eyes are postulated for the larvae in the ground pattern of the Eumollusca. These eyes are composed of pigment cells with microvilli and rhabdomeric sensory cells with cilia. Both cells contact the outer medium apically. The eyes are situated caudally to the prototroch and innervated by the lateral nerve cords. Cerebral innervation must have evolved at least in the stem lineage of the Ganglioneura and the modification of the position of the eyes from posttrochal to praetrochal must at least have taken place in the stem lineage of the Gastropoda. According to these considerations the larval eyes must have been reduced secondarily in the Scaphopoda and some Bivalvia.

In the larvae of Solenogastres and Caudofoveata no eyes have been found until now. The evaluation of the absence of eyes in larva of both taxa still seems to be a point of controversial discussion (ROSEN et al. 1978, 1979 versus SALVINI-PLAWEN 1982, 1988). Two alternative hypotheses seem to be possible:

– The larva in the ground pattern of molluscs had no eyes. The absence of

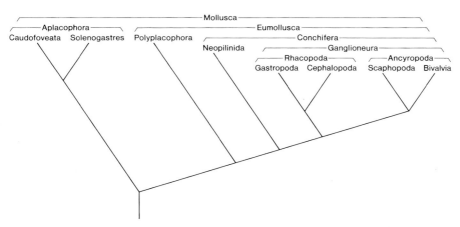

Fig. 13: Diagramm of the phylogenetic relationships within the Mollusca modified from LAUTERBACH (1986).

photoreceptors in the larvae of Solenogastres and Caudofoveata then would be plesiomorphous.

– If larvae, on the other hand, were homologous within Bilateria with a spiral cleavage or even if eyes generally appeared early in development in this group, the most parsimonious explanation would be that the larva in the ground pattern of molluscs already had eyes. The eyes in the larvae of Solenogastres and Caudofoveata then would have been reduced in their common stem lineage, so that the absence of larval eyes would be an autapomorphy of the Aplacophora.

At the present stage of investigation the latter hypothesis seems to be more likely.

Zusammenfassung

Die pelagische Larve von *Lepidochiton cinereus* besitzt ein Paar dorsal gelegener intraepidermaler Augen. Die Augen sind evers und bestehen aus vier langgestreckten Rezeptorzellen und sechs hochprismatischen Pigmentzellen. Beide Zellformen sind mit zwei bis drei Cilien bewimpert; die Zelloberfläche weist außerdem zahlreiche Mikrovilli auf. Im Verlauf der Ausbildung von Schalenplatten werden die Augen ventrad verschoben und von der Epidermis überwachsen. Dabei kommt es zu einer Stauchung der Photorezeptoren. Postlarvale, bodenlebende mit acht dorsalen Schalenplatten ausgestattete *Lepidochiton cinereus* besitzen daher ein Paar ventrolateral gelegener, becherförmiger Augen, deren Pigment- und Rezeptorzellen intensiv verzahnt sind. Die apikale Zelloberfläche beider Zellformen weist Cilien und stark verlängerte Mikrovilli auf, die das Zentrum des Augenbechers vollständig ausfüllen. Die Veliger-Larve von *Lacuna divaricata* besitzt ein Paar subepidermaler Photorezeptoren, die aus einer Pigmentzelle, zwei unterschiedlichen Rezeptorzellen und einer Cornea-Zelle bestehen, die eine Linse sezerniert. Die Perzeption der optischen Information erfolgt über eine rhabdomerische Rezeptorzelle, die zahlreiche langgestreckte und gerade Mikrovilli sowie ein Cilium aufweist, und über eine ciliäre Rezeptorzelle, die zahlreiche Cilien und kurze unregelmäßig geformte Mikrovilli besitzt. Ein Vergleich innerhalb der Mollusken zeigt, daß die Larve im Grundmuster der Eumollusken bereits ein Paar intraepidermaler Augenflecken besessen haben mußte, die aus mehreren Pigmentzellen und rhabdomerischen Rezeptorzellen bestanden und keine Linse besaß. Demnach wurden ciliäre Rezeptorzellen sowie eine Linse erst innerhalb der Gastropoden evolviert.

Literature

ARNOLD, J. M. (1966): On the occurence of microtubules in the developing lens of the squid *Loligo pealii.* J. ultrastr. Res. **14**, 534–539.
BARTOLOMAEUS, T. (1989): Larvale Nierenorgane bei *Lepidochiton cinereus* (Polyplacophora) und *Aeolidia papilloosa* (Gastropoda). Zoomorphology **108**, 297–307.
BERTHOLD, T. & T. ENGESER (1987): Phylogenetic analysis and systematization of the Cephalopoda (Mollusca). Verh. naturwiss. Ver. Hamburg **29**, 187–220.
BOYLE, P. R. (1969): Fine structure of the eyes of *Onithochiton neglectus* (Mollusca: Polyplacoophora). Z. Zellforsch. **102**, 313–332.
BUCHANAN, J. A. (1986): Ultrastructure of the larval eyes of *Hermissenda crassicornis* (Mollusca, Nudibranchia). J. ultrastr. molec. struc. Res. **94**, 52–62.
CHIA, F.-S. & R. KOSS (1983): Fine structure of the larval eyes of *Rostanga pulchra* (Mollusca, Opisthobranchia, Nudibranchia). Zoomorphology **102**, 1–10.
CROWTHER, R. J. & D. B. BONAR (1980): Photoreceptor ontogeny and differentiation in *Ilyanassa obsoleta.* Am. Zool. **20**, 891.
EAKIN, R. M. & J. L. BRANDENBURGER (1967): Differentiation in the eye of a pulmonate snail *Helix aspersa.* J. ultrastr. Res. **18**, 391–421.
EAKIN, R. M. & J. A. WESTFALL (1964): Further observations on the fine structure of some invertebrates eyes. Z. Zellforsch. **62**, 310–332.
EAKIN, R. M., J. A. WESTFALL & M. J. DENNIS (1967): Fine structure of the eye of a nudibranch mollusc, *Hermissenda crassicornis.* J. Cell Sci. **2**, 349–358.
EAKIN, R. M., J. L. BRANDENBURGER & G. M. BARKER (1980): Fine structure of the eye of the New Zealand slug *Athoracophorus bitentaculatus.* Zoomorphologie **94**, 225–239.
FISCHER, F. P. (1978): Photoreceptor in chiton aesthetes. Spixiana **1**, 209–213.
FISCHER, F. P. (1979): Die Aestheten von *Acanthochiton fascicularis* (Mollusca, Polyplacophora). Zoomorphologie **92**, 95–106.
FISCHER, F. P. (1980): Fine structure of the larval eye of *Lepidochiton cinerea* L. (Mollusca, Polyplacophora). Spixiana **3**, 53–57.
FRETTER, V. (1969): Aspects of metamorphosis in prosobranch gastropods. Proc. Malac. Soc. Lond. **38**, 375–386.
GIBSON, B. L. (1984): Cellular and ultrastructural features of the adult and the embryonic eye in the marine gastropode, *Ilyanassa obsoleta.* J. Morphol. **181**, 205–220.
HASZPRUNAR, G. (1988): On the origin and evolution of major gastropod groups, with special reference to the Streptoneura. J. Moll. Stud. **54**, 367–441.
HUGHES, H. P. I. (1970 a): A light and electron microscope study of some opisthobranch eyes. Z. Zellforsch. **106**, 79–98.
HUGHES, H. P. I. (1970 b): The larval eye of the aeolid nudibranch *Trinchesia aurantia* (Alder and Hancock). Z. Zellforsch. **109**, 55–63.
LAUTERBACH, K.-E. (1983): Erörterungen zur Stammesgeschichte der Mollusca, insbesondere der Conchifera. Z. zool. Syst. Evolutionsforsch. **21**, 201–216.
MOOR, B. (1983): Organogenesis. In: The Mollusca. Vol. 3, Development. (Eds.: N. H. VERDENK, J. A. M. VAN DEN BIGGELAAR, A. S. TEMPA). Academic Press, New York, London. P. 123–177.
RAVEN, C. P. (1958): Morphogenesis: The analysis of the mollscan development. Pergamon Press, London. 311 p.
ROSEN, M. D., C. R. STASEK & C. O. HERMANS (1978): The ultrastructure and evolutionary significance of the cerebral ocelli of *Mytilus edulis,* the Bay Mussel. The Veliger **21**, 10–18.
ROSEN, M. D., C. R. STASEK & C. O. HERMANS (1979): The ultrastructure and evolutionary significance of the ocelli in the larva of *Katharina tunicata.* The Veliger **22**, 173–178.
SALVINI-PLAWEN, L. v. (1982): On the polyphyletic origin of Photoreceptors. In: Visual cells in Evolution (ed: J. A. WESTFALL). Raven Press, New York. P. 137–154.
SALVINI-PLAWEN, L. v. (1988): Annelida and Mollusca – a prospectus. In: The ultrastructure of the Polychaeta (eds: W. WESTHEIDE & C. O. HERMANS). Microfauna Marina **4**, 383–396.

SMITH, F. G. W. (1935): On the development of *Patella vulgata.* Phil. Trans. R. Soc. London B **225**, 95–125.
TONOSAKI, A. (1965): The fine structure of the retinal plexus in *Octopus vulgaris.* Z. Zellforsch. **79**, 521–532.
TONOSAKI, A. (1967): The fine structure of the retina of *Haliotis discus.* Z. Zellforsch. **79**, 469–480.

Dr. Thomas Bartolomaeus
II. Zoologisches Institut und Museum der Universität Göttingen
Berliner Straße 28, D-3400 Göttingen

On the ultrastructure of the cuticle, the epidermis and the gills of *Sternaspis scutata* (Annelida)

Thomas Bartolomaeus

Contents

Abstract	237
A. Introduction	238
B. Material and Methods	239
C. Results	239
Cuticle, epidermis and ventro-caudal shield	239
Gills	244
D. Discussion	248
Zusammenfassung	250
Literature	251

Abstract

Cuticle, parts of the epidermis, the ventro-caudal shield and the gills of *Sternaspis scutata* have been investigated ultrastructurally. The cuticle is composed of electron-bright fibers which are partly arranged upon an orthogonal pattern and which are embedded in an amorphous matrix. The cuticle varies between 75 and 150 µm in thickness and is formed by a cuboidal monolayer of epidermal supporting cells, which contain a prominent tonofilament system. The ventro-caudal shield consists of an extremely hard electron-black material. A meshwork of fibers provides a framework for this material. According to histochemical tests, this material is mineralized iron. A large number of helical gills extend from the perianal region into the surrounding medium. These gills consist of a monolayer of epidermal cells which are covered by a cuticle. Inside, there are two parallel canals each of which is bordered by a monolayer of muscle cells. These rest either on the subepidermal ECM or on a transverse band of ECM, which is continuous with the subepidermal ECM. Blood vessels, which are bordered by ECM only, can be found within the transverse band or in the subepidermal ECM.

A. Introduction

Sternaspis scutata is a cosmopolitan sublitoral species, which lives in muddy sediments (PETTIBONE 1954, HARTMANN-SCHRÖDER 1971). There are only a few studies on the ultrastructure of this aberrant animal, which has an unknown systematic position within polychaete annelids. These studies deal with observations on the oogenesis and the egg (NOTARBARTOLO 1970, VILLA 1976) and with scanning electron microscopy of the epidermis (RALLO & MOYA 1987).

The most striking feature of *Sternaspis scutata* is one pair of ventral shield-like plates which covers the caudal fourth of the animal (Fig. 1). Bundles of chaetae

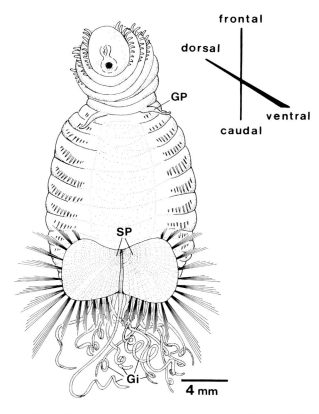

Fig. 1: *Sternaspis scutata*. Ventral view, according to own observations and RIEDL (1983). The anterior 6 segments can be retracted into more posterior ones, segments 1 and 5–7 are achaetous. The paired genital papilla (GP) can be seen in segment 7. Segments 8–14 midventrally without segmental divisions and lateral intraintegumental chaetae. Ventrally, the posterior segments are covered by a pair of shield-like plates (SP) each with 16 radiating bundles of chaetae. A very large number (much more than drawn here) of filiform gills (Gi) can be seen caudally.

radiate from the shield. This paper deals with the ultrastructure of these shield and compares them to the dorsal cuticle. In addition, this the caudally situated filiform gill appendages are described ultrastructurally.

This investigation was financially supported by the Akademie der Wissenschaften und der Literatur, Mainz. I would like to thank Dr. T. Kunert for kindly correcting the manuscript, Mr. G. Görrissen for collecting the specimen and Mr. M. Uhr for advices on histochemical tests.

B. Material and Methods

Sternaspis scutata Ranzani, 1817 was collected from muddy sediments off Marseille close to the isle of Château d'If and kept alive for 12 h at 4° C at the Station Biologique de la Tour du Valat, Le Sambuc. The animal was dissected in a fixative consisting of 2.5 % glutaraldehyde buffered in 0.1 M sodium cacodylate at 4° C. After 10 h of fixation the specimen was washed four times in 0.1 M sodium-cacodylate buffer and kept in the same buffer for one week. After 90 min of postfixation in a sodium cacodylate buffered 1 % OsO_4-solution, pieces of *S. scutata* were dehydrated in an acetone series and embedded in araldite. 75 to 85 nm sections were cut on glas and diamond knifes with a REICHERT Ultracut microtome. The sections were kept on formvar-covered single slot copper grids, automatically strained in a LKB Ultrastainer and examined with a ZEISS EM 10 B electron microscope. It was difficult to section the ventrocaudal shield, because the material is so hard that the razor blades used for preparing the araldite-embedded material for ultramicrotomy twist soon after they have penetrated the shield. Sections of the ventro-caudal shield and its lateral margins were compared to the dorsal cuticle and the posterior chaetae and were tested for iron by the Berlinerblau-reaction (GOMORI 1936). Complete series of sections were made of parts of the gill filaments.

C. Results

Cuticle, epidermis and ventro-caudal shield

The dorsal cuticle measures between 75 and 90 µm, sometimes the cuticle is much higher (up to 150 µm laterally to the shield). The cuticle consists of an amorphous electron-grey matrix which is crossed by unbanded, translucent hollow-fibers. These structures enlarge in diameter across a baso-apical gradient and seem to contain small electron-grey invaginations from the surrounding matrix (Fig. 2 B). The layers of fibers are not always arranged upon an orthogonal pattern. This seems to be characteristic of the cuticle close to the ventro-caudal shield and its apical layers. More basally the dorsal cuticle contains electron-dense rods with electron-dark spots inside. These rods have a specific external structure with ringlike constrictions and small protrusions. They seem to be synthesized intraepidermally and are directly discharged into the cuticle (Fig. 2 A). Small fibrils are situated between and apically to the microvilli of the supporting cell and seem to be orthogonally arranged. More apically such fibrils can not be found.

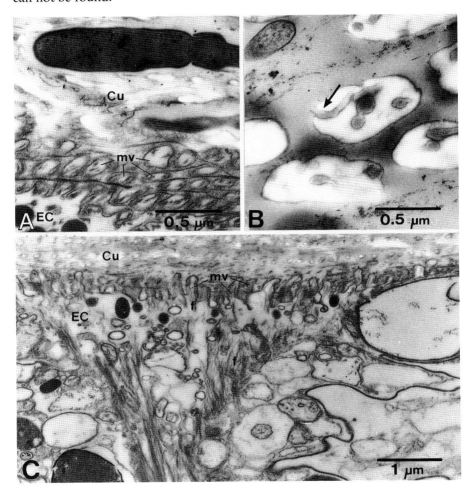

Fig. 2: Cuticle and epidermal supporting cells (EC). A. Fibers and electron-dense rods can be seen basally in the cuticle (Cu) close to the microvilli (mv). B. Translucent hollowfibers with electron-grey invaginations (arrow) cross the cuticle. C. Supporting cells with tonofilaments (f) extend into the microvilli.

Close to the ventro-caudal shield the proportion of fibrils in the basal regions of the cuticle drastically increases. The fibrils, which measure about 40 nm in diameter, are interwoven one into another. They are composed of several smaller filaments and form a kind of meshwork, which surrounds the microvilli of the supporting cells. This net of fibrils seems to provide the framework for the electron-black material (iron) the shield is composed of (Fig. 3 A).

The epidermis mainly consists of supporting cells, which are involved in the formation of the cuticle. No gland or sensory cells have been observed. The latter are expected to be more frequent in the frontal parts of the animal. In the

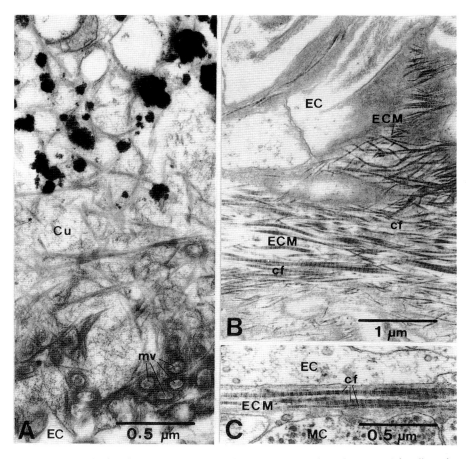

Fig. 3: Cuticle and subepidermal ECM (extracellular matrix). A. Basal cuticle (Cu) peripherally to the ventro-caudal plate. Note the netlike arrangement of fibers apically to the epidermal microvilli (mv). B. The subepidermal ECM consists of collagen fibers (cf) embedded in amorphous material. C. Collagen fibers. EC = supporting cell, MC = muscle cell.

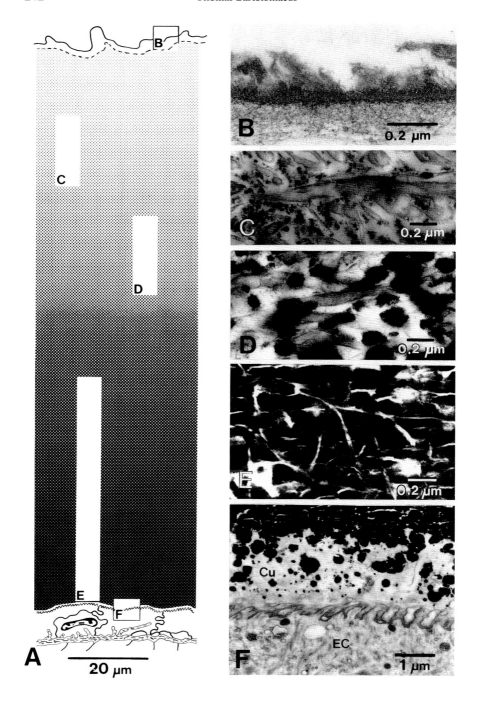

regions investigated supporting cells form a monolayer of about 6 to 10 μm thickness. The supporting cells interdigitate intensely and bear numerous short microvilli apically, which measure 0.5 to 0.75 μm in length and 120 nm in diameter. Solely in the regions of the ventro-caudal shield, some of the microvilli are longer and extend deeply into the cuticle. Their tips are swollen, so that the cuticle seems to contain vesicles (Fig. 3 A). The supporting cells are connected with one another by belt desmosomes and septate junctions. Generally the nucleus is situated medially and may be irregularly shaped. The cells contain vacuoles, lysosomes, mitochondria situated subapically, several vesicles of different sizes and contents as well as rough endoplasmic reticulum. Vesicles with electron dense contents which measure 170 nm in diameter can only be found directly underneath the cells apex. No coated vesicles and coated pits have been found.

The most remarkable feature of the supporting cells is a meshwork of tonofilaments, which mainly cross the cell in an baso-apical direction and extend into the microvilli (Fig. 2 C). Here, they are seen to be attached to the lateral margins and not to the tip. Basally, bundles of tonofilaments are attached to the subepidermal extracellular matrix (ECM) by hemidesmosomes. Between the bases of the supporting cells one regularly finds clusters of neurites.

The subepidermal ECM consists of three layers, i. e. two layers of amorphous electron-grey material enclose a layer of banded collagen fibers (Fig. 3 B, C). This layer ranges between 1.5 μm and less than 250 nm in thickness. The ECM may extend into basal infoldings of the supporting cells.

Electron-black material is clustered between the fibrillar meshwork of the cuticle which constitutes the ventro-caudal shield (Fig. 4 E, F). In the apical one third, the concentration of electron-black material decreases. Spots of this material, which become smaller towards the apex can be seen between hollow-fibers (Fig. 4 C, D). These fibers often bear small lateral branches and are above arranged in orthogonal layers as mentioned. They measure between 40 nm and 140 nm. 6 to 8 μm below the epicuticle there are no electron-black spots and 1 μm below the epicuticle there are no fibers. The epicuticle consists of an 45 to 60 nm thin electron-dense layer. Several tightly packed electron-grey protuberances originate from this originate from this layer. These are not membrane-bound. It seems likely that the electron-black material the ventro-caudal shield

◄ Fig. 4: Ventro-caudal shield. A. Schematical cross section to illustrate different electron-density within the plate. B.–F. Large magnifications from regions indicated in A. B. Epicuticle. C. Small electron-dense spots between the fibers. D. Spots increase in size. E. Fiber cross nearly compact electron-dense material. F. Supporting cells (EC) and basal cuticle (Cu) with clustering electron-dense spots.

consists of mineralized iron. This is substantiated in the Berlinerbau-reaction. Those regions which consist of the electron-dense material turn intensely blue and thus it is proved that they contain iron. Beside this, those parts of the chaetae, which contain the same electron-black particles (Fig. 9) also turn blue in the Berlinerbau-reaction. No reaction was obtained in the dorsal cuticle and the cuticle adjacent to the shield.

Gills

A large number of long, filiform and sometimes spiralled gills extends from the caudal end of the animal into the medium. In living animals these gills are

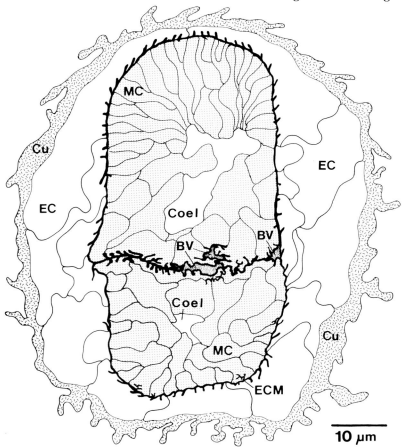

Fig. 5: Cross section of a gill filament. Internally, the filament consists of two parts, each of which contains a small coelomic cavity (Coel). Muscle cells (MC) rest on extracellular matrix (ECM) which dilates to form blood vessels (BV). Cu = cuticle, EC = supporting cell.

moved slowly. They seem to form spirals and contract cork-screw like, then they relax, elongate and contract again.

Cross sections reveal the organization of the gill filament (Fig. 5). The gills are covered by a cuticle which is followed by a monolayer of epidermal cells. These rest on the subepidermal ECM. Two intraepidermal nerves are found. Inside, there are two neighbouring monolayers of muscle cells, which are completely separated from each other by a transversal band of ECM. This material is continuous with the subepidermal ECM. The monolayers of muscle cells each border a small cavity, which is presumed to be of coelomic origin. It seems likely that they are in contact with the coelom of the last segments. The main blood vessel can be seen in the middle of the transverse band of ECM. Smaller lateral blood vessels are found in the subepidermal ECM especially close to the transverse band (Fig. 5).

The cuticle consists of an amorphous electron-grey material, which surrounds irregularly arranged electron-dense fibers measuring 35 nm in diameter. The cuticle measures 1 µm to 3.5 µm in thickness and forms large protuberances which extend up to 7.5 µm into the surrounding medium (Fig. 6). They seem to

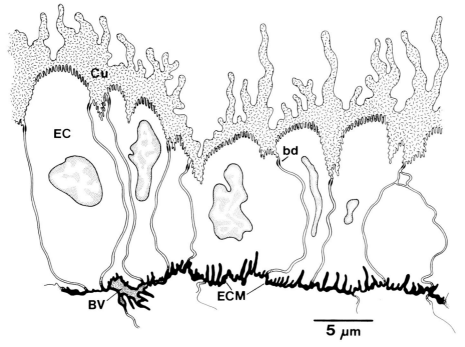

Fig. 6: Cuticle (Cu) and epidermis of the gill filaments. The subepidermal extracellular matrix (ECM) dilates and forms blood vessels (BV). bd = belt desmosome, EC = supporting cell.

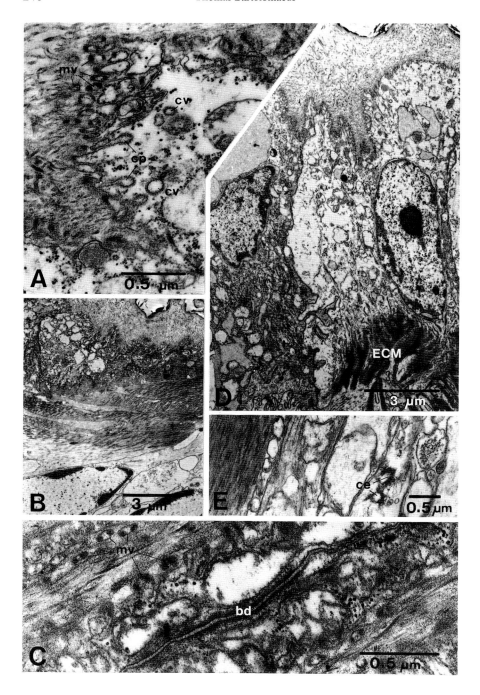

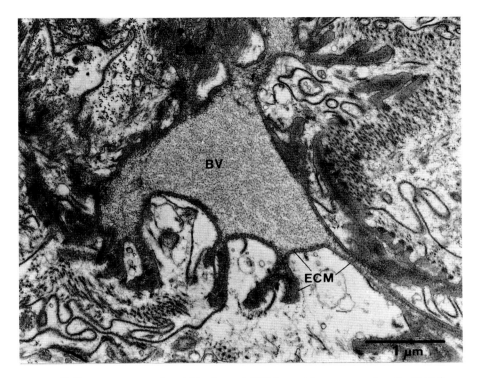

Fig. 8: Gill filament. Cross sectioned blood vessel (BV) located between muscle cells. Fluid flows between amorphous extracellular material (ECM) which may contain small fibers (f).

indicate the great flexibility of the cuticle, which has to withstand contractions of the gill muscles.

The epidermis consists of unciliated supporting cells, which are likely to secrete the cuticle (Fig. 6, 7 D). No gland cells or sensory cells have been found in the region investigated. The supporting cells form a monolayer, interdigitate and are connected by apical belt desmosomes and septate junctions. Short microvilli extend from the apical surface into the cuticle. Cross-sectioned microvilli have an oval shape; each microvillus contains two pronounced lateral groups of electron-dense structures (Fig. 7 C). Between these microvilli, coated pits extend deeply into the cell body. They are related to coated vesicles situated

◄ Fig. 7: Gill filament. A. Supporting cell with coated pits (cp) and coated vesicles (cv). B. Cuticle, epidermis and muscle cells. C. Sagittal section of the supporting cell. Note two lateral electron-dense cores inside the microvilli. D. Epidermal cells and intensely folded extracellular matrix (ECM). E. Muscle cell with a pair of centrioles (ce). bd = belt desmosome.

subapically, which have a clear contents with some flocculent material close to the vesicle membrane (Fig. 7 A). Beside the medially situated nucleus, which may be irregularly shaped, the cells contain vacuoles with electron-grey material, mitochondria, lysosomes, rough endoplasmic reticulum and vesicles of different sizes. Some of the supporting cells in the gills have a much more electron-dense cytoplasm compared to neighbouring cells, but beside this no structural differences have been detected (Fig. 7 D). The supporting cells have no or solely a weak system of tonofilaments. Thus, they can be deformed by the mechanical forces of the contracting muscle cells. This leads to an irregular shape of the supporting cells.

Opposite the epidermal cells muscle cells rest on the subepidermal ECM (Fig. 7 B). The myofibrils are concentrated in the basal regions of the cell, while the nucleus is situated more apically. Most of the myofibrils are arranged in a longitudinal direction. Circular muscle cells are solely found close to the blood vessels. Generally a pair of centrioles is situated underneath the apical cell membrane and a short rootlet may be attached to one of both (Fig. 7 E). According to the state of contraction, the coelomic cavity may be very small.

The ECM consists of amorphous electron-grey material which seldom contains very small electron-dense fibrils. The ECM borders the blood vessels, the largest of which is situated in the transversal band. No special blood cells have been detected inside the vessels, solely electron-grey particles are precipitated inside the vessels (Fig. 8). Hypotheses on the function of the gills will be discussed below.

D. Discussion

A tonofilament system running from the cell apex to the base has been observed in several annelid taxa (STORCH & WELSCH 1970, WELSCH et al. 1984, STORCH 1988). They are supposed to be a predominant feature in cells subjected to mechanical stress. In *Sternaspis scutata* this seems to be true only for those supporting cells which mediate mechanical force from muscle cells to an inflexible cuticle. The thick cuticle of the body wall and the ventro-caudal shield cannot be deformed by subepidermal muscles. Here, the cuticle functions as an inflexible insertion structure for the muscles, and the supporting cells are stabilized by a tonofilament system. Accordingly, such a system lacks in the supporting cells of the gills. Here, the epidermal cells and the cuticle are deformed by strong subepidermal muscles, which leads to a spiralling of the gills. Tonofilaments can only be found inside the microvilli. Here, they are restricted to a pair of laterally situated electron-dense elements, which are but groups of axial tonofilaments (RICHARDS 1984, p. 311).

A cuticle consisting of orthogonal layers of unbanded collagen fibers which are embedded in an amorphous matrix are reported not only from the Annelida (see RICHARDS 1984, STORCH 1988), but also from the Pogonophora (SOUTHWARD 1984), Sipunculida (MORITZ & STORCH 1970) and Echiurida (own observations). Such a cuticle must therefore represent a plesiomorphic character of annelids. An orthogonal arrangement of the fibers has only been observed in subapical cuticular material of the shield. Here, the fibers are branched additionally. In the dorsal and ventral cuticle investigated so far, the fibers are oriented irregularly. From the basis of the cuticle to the apex they seem to increase in diameter. This swelling is most likely to be genuine. This modification in the orientation and the structure of the fibers is with much probability apomorphic in *Sternaspis scutata*. In most representatives of the mentioned taxa, i. e. Annelida, Pogonophora, Sipunculida, microvilli originating from the supporting cells extend deeply into the cuticle or traverse the cuticle. Exceptions like that observed in *Sternaspis scutata* have been reported also for the cirri and elytra of *Gastrolepidia clavigera* and for the polychaete *Arabella irricolor* (STORCH 1988). The absence of transcuticular epidermal microvilli is said to be correlated to a protective function of the cuticle (STORCH 1988).

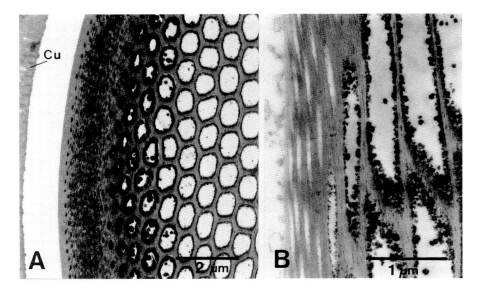

Fig. 9: Caudal chaetae inside the follicle. A. Cross section, close to the exterior. B. Longitudinal section, close to the chaetoblast. The cuticle (Cu) extends deeply into the follicle. Note the electron-dense spots inside the chaetae.

The most remarkable cuticular structure is the paired ventro-caudal shield. In alcohol it has a rusty red colour (PETTIBONE 1954). The material the shield consists of is so hard that it destroys razor blades when being trimmed for ultramicrotomy. In the electron microscope it appears black, even in unstained sections. The same material can be seen inside the chaetae (Fig. 9). A comparable material has been described in the radular teeth of *Cryptochiton stelleri* as mineralized iron (TOWE & LOWENSTAM 1967). For this species, a three-dimensional meshwork of fibers consisting of smaller fibrils is described as to represent the framework for the iron particles. The metal first appears as small spots which increase and aggregate to larger assemblages. These observations remarkably correspond to the situation found in *Sternaspis scutata*. A histochemical test for iron, i. e. the Berlinerblau-reaction, is positive in those areas which are of the described electron-black appearance in the electron microscope. It therefore is presumed that mineralized iron, is the main chemical component of the shield. A chemical microanalysis would be helpful for a final clarification of the chemical component iron is bound to.

Gills have been investigated ultrastructurally in different annelid taxa (STORCH & ALBERTI 1978, MENENDEZ et al. 1984, STORCH & GAILL 1986, GARDINER 1988). Only some of these have gills with blood vessels surrounded by muscle cells like in *Sternaspis scutata*. The few gills that have been investigated in polychaetes show a great anatomical diverity and thus, at the moment no evolutionary aspects on these gills can be discussed. Nevertheless, a few points of the functions of the gills in *Sternaspis scutata* should be mentioned. As the blood spaces are located in the subepidermal ECM, oxygen has to diffuse across the cuticle and the epidermal cells. In the living animal it has been observed that the filiform gills are moved slowly; they become spiralled, elongated and spiralled again. This movement is caused by an alternating contractions of muscles in one half of the gills. As the muscle-fibers are mostly oriented longitudinally, the pressure inside the two coelomic cavities acts antagonistically to the muscles. Repeated helical movements of the gills might keep the blood fluid in motion.

Zusammenfassung

Kutikula, Epidermis, die caudoventralen Schildplatten und die Kiemen wurden ultrastrukturell untersucht. Die Kutikula besteht aus nicht-orthogonal angeordneten elektronenhellen Fibrillen, die in eine amorphe Grundsubstanz eingebettet sind. Die Kutikula ist 75 bis 150 µm stark. Sie wird von einer einschichtigen kubischen Epidermis gebildet, deren zelluläre Hauptkomponente aciliäre Trägerzellen darstellen. Diese Zellen weisen ein auffälliges Tonofilament-System auf. Die paarigen caudoventralen Schildplatten bestehen aus sehr hartem, elektronendichtem Material. Das Grundgerüst für die Einlagerung dieses Materials stellt ein Netzwerk dar, dessen Fasern aus feinen Fibrillen zusammengesetzt sind. Mit Hilfe der Berlinerblau-Reaktion läßt sich nachweisen, daß es sich bei dem elektronendichten Material um Eisen handelt. Die fadenförmigen, spiraligen Kiemen von *Sternaspis scutata* entspringen der perianalen Region. Sie bestehen aus einer einschichtigen Epidermis, der eine Kutikula aufliegt und die in einer elektronendichten extracellulären Matrix (ECM) fußt. Im Innern der Kieme verlaufen benachbart zwei Kanäle, die von einer einschichtigen Lage von Muskelzellen umgeben sind. Diese Zellen ruhen auf der subepidermalen ECM oder auf einem mit dieser in Verbindung stehenden transversalen Band von ECM. Nicht-endothelial begrenzte Blutgefäße verlaufen in dem transversalen Band oder im Übergangsbereich zur subepidermalen ECM.

Literature

GARDINER, S. L. (1988): Respiratory and feeding appendages. In: The ultrastructure of the Polychaeta (eds.: W. WESTHEIDE & C. O. HERMANS). Microfauna Marina **4**, 37–43.

GOMORI, G. (1936): Microtechnical demonstration of iron. Am. J. Pathol. **12**, 655–663.

HARTMANN-SCHRÖDER, G. (1971): Annelida, Polychaeta. In: Die Tierwelt Deutschlands und der angrenzenden Meeresteile nach ihren Merkmalen und nach ihrer Lebensweise. 58. Teil. (eds.: M. DAHL & F. PEUS). G. Fischer Verlag, Jena.

MENENDEZ, A., J. L. ARIAS, D. TOLIVIA & A. ALVAREZ-URIA (1984): Ultrastructure of gill epithelial cells of *Diopatra neapolitana* (Annelida, Polychaeta). Zoomorphology **104**, 304–309.

MORITZ, K. & V. STORCH (1970): Über den Aufbau des Integumentes der Priapuliden und der Sipunculiden [*Priapulus caudatus* Lamarck, *Phascolion strombi* (Montagu)]. Z. Zellforsch. **105**, 55–64.

NOTARBARTOLO, L. (1970): Morfologia ultrastrutturale degli involucri corticale e del nucleo dell'uovo ovarico maturo de *Sternaspis*. Acta Embryol. Exp., 133–150.

PETTIBONE, M. H. (1954): Marine polychaete worms from Point Barrow, Alaska, with additional records from the North Atlantic and North Pacific. Proc. U. S. Nat. Mus. **103**, 203–356.

RALLO, A. & J. MOYA (1987): Detalles de la anatomica externa de *Sternaspis scutata* (Annelida: Polychaeta), estudiada con microscopia electronica de barrido. Actas VIII Bienal. R. Soc. Esp. Hist. Nat. Pamplona **94**, 105–112.

RICHARDS, K. S. (1984): Cuticle. In: Biology of the integument (eds.: J. BEREITER-HAHN, A. G. MATOLTSY & K. S. RICHARDS). Springer Verlag, Berlin. P. 310–322.

RIEDL, R. (1983): Fauna und Flora des Mittelmeeres. 3. Auflage. P. Parey, Hamburg, Berlin. 836 p.

SOUTHWARD, E. C. (1984): Pogonophora. In: Biology of the integument (eds: J. BEREITER-HAHN, A. G. MATOLTSY & K. S. RICHARDS). Springer Verlag, Berlin. P. 376–388.

STORCH, V. (1988): Integument. In: the ultrastructure of the Polychaeta (eds: W. WESTHEIDE & C. O. HERMANS). Microfauna Marina **4**, 13–36.

STORCH, V. & F. GAILL (1986): Ultrastructural observations on feeding appendages and gills of *Alvinella pompejana* (Annelida, Polychaeta). Helgoländer Meeresunters. **40**, 309–319.

STORCH, V. & G. ALBERTI (1977): Ultrastructural observations on the gills of polychaetes. Helgoländer wiss. Meeresunters. **31**, 169–179.

STORCH, V. & U. WELSCH (1970): Über die Feinstruktur der Polychaeten-Epidermis (Annelida). Z. Morph. Tiere **66**, 310–322.

TOWE, K. M. & LOWENSTAM, H. A. (1976): Ultrastructure and development of iron mineralization in the radular teeth of *Cryptochiton stelleri* (Mollusca). J. ultrastr. Res. **17**, 1–13.

VILLA, L. (1976): An ultrastructural investigation of the polar plasm of the egg of *Sternaspis* (Annelida, Polychaeta). Acta. Embryol. Exp. **2**, 153–165.

WELSCH, U., V. STORCH & K. S. RICHARDS (1984): Epidermal cells. In: Biology of the integument (eds: J. BEREITER-HAHN, A. G. MATOLTSY & K. S. RICHARDS). Springer Verlag, Berlin. P. 269–296.

Dr. Thomas Bartolomaeus
II. Zoologisches Institut und Museum der Universität Göttingen
Berliner Straße 28, D-3400 Göttingen

Dermonephridia – modified epidermal cells with a probable excretory function in *Paratomella rubra* (Acoela, Plathelminthes)

Ulrich Ehlers

Contents

Abstract	253
A. Introduction	254
B. Materials and Methods	254
C. Results	254
D. Discussion	260
Acknowledgements	262
Zusammenfassung	262
Abbreviations	263
References	263

Abstract

Paratomella rubra possesses several specialized epidermal cells in different body regions. These unlinked cells, which are called dermonephridia, are unciliated, their peripheral cytoplasm is disintegrated in many thread-like projections resembling slender microvilli. Most parts of the cells are filled with a highly branching extracellular lacunar system of interconnecting tubules and vacuoles and with smaller isolated vesicles. The basal regions of these cells border cells of the central digestive system and of the peripheral "parenchymal" digestive system as well as other cell types like gland cells. An excretory function for the modified epidermal cells is discussed.

An explanation is given for the fact, that *Paratomella* like all other representatives of the Acoelomorpha lacks any protonephridia. The dermonephridia are hypothesized as an autapomorphy of the taxon *Paratomella*.

A. Introduction

Within the Plathelminthes, protonephridia are present in all taxa except the Nemertodermatida and the Acoela (EHLERS 1985, 1989; RIEGER et al. 1991). Up to now, any organs or specialized cells with an excretory function have been unknown in any species of these two taxa, which together constitute the monophylum Acoelomorpha.

During the course of EM-investigations of the acoel *Paratomella rubra* Rieger & Ott, 1971, a conspicuous type of strongly modified epidermal cells was found. Such cells are unknown for any other taxon of the Plathelminthes or the Bilateria in general. The specialized organization of these cells gives rise to the conclusion of an excretory function.

B. Materials and Methods

The material investigated belongs to the East Atlantic population of the amphiatlantic species *Paratomella rubra* Rieger & Ott, 1971 (RIEGER & OTT 1971; CREZEE 1978; EHLERS & EHLERS 1980; EHLERS 1992 a). Specimens were extracted from sandy sediments taken at Gran Canaria in 1979 and 1980, respectively. For electron microscopical observations unanaesthetized mature animals were fixed in 2.5 % glutaraldehyd in 0.1 M sodium cacodylate buffer (pH 7.2–7.4) at 4° C for 2 h, rinsed in the same buffer, postfixed in 1 % osmium tetroxide in the same buffer for 1 h, dehydrated in an acetone series and embedded in Araldite. Serial ultrathin sections were stained with aqueous uranyl acetate and lead citrate and examined with ZEISS EM 10 B and ZEISS EM 900 electron microscopes.

C. Results

The cellular epidermis of *Paratomella rubra* is not markedly different from that of many other acoels (Figs. 1, 2). A subepidermal basal lamina and a reticular lamina are entirely lacking. Thin circular and stronger longitudinal muscles and neurons are interposed between the epidermal cells. The epidermal nuclei lie below the body wall muscles, i. e., the epidermis is of the "insunk" type.

Cilium-bearing sensory dendrites and outleading necks of different types of subepidermal glands protrude between the surface portions of the epidermal cells.

Besides this typical and well known organization, epidermal cells with a special appearance have been found in different body regions, e. g. near the

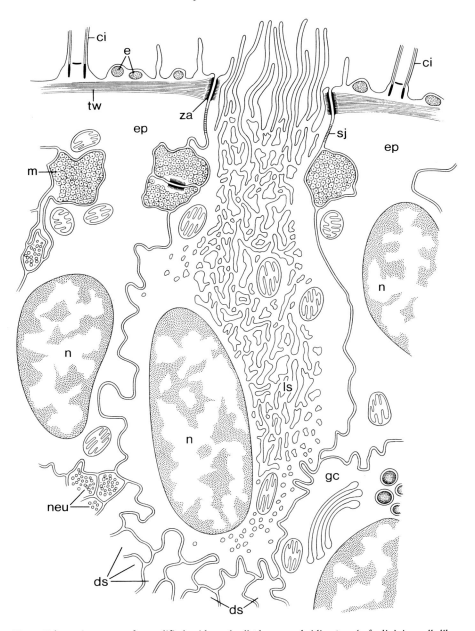

Fig. 1- Schematic pattern of a modified epidermal cell (dermonephridium) and of adjoining cells like ciliated epidermis cells, muscle cells, gland cells and cytoplasmic processes of nerve cells and cells of the digestive system.

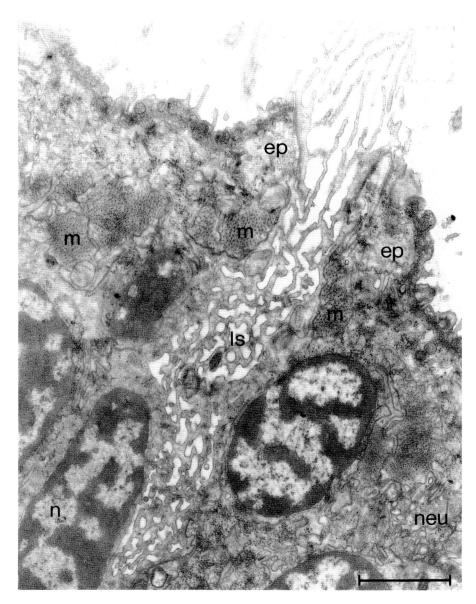

Fig. 2. Section through the more peripheral parts of a dermonephridium with its extracellular lacunar system and the distal thread-like cytoplasmic projections. X 24 000. Scale bars in Figs. 2–4: 1 µm.

statocyst, in the vicinity of the mouth, and near the follicular testes, respectively. The unlinked specialized epidermal cells are not arranged in a bilaterally symmetrical or serial manner, but seem to be dispersed randomly.

These cells show a relatively small surface portion in which cilia and epitheliosomes are absent. The peripheral cytoplasm is disintegrated in many thread-like projections resembling long and slender microvilli (Figs. 1, 2, 4 A). Most often, these parts of the cells protrude above the surface of the neighbouring unmodified and ciliated epidermal cells.

The lumina between the thread-like projections continuate more proximally into a prominent and highly branched lacunar system which fills most parts of the specialized cell (Figs. 1–4). The system is of an irregular appearance and consists of many communicating tubules and vacuoles of varying diameter. This interconnected extracellular system extends deeply into the cell body of the modified epidermal cell; most often the nuclei of these cells are more "insunk" than those of bordering unmodified epidermal cells (Figs. 1–3). In addition, numerous small vesicles have been found in the cytoplasm of the modified cells, especially in its basal parts where the lacunar system seems to originate (Figs. 1, 3). Besides the lacunar system and the vesicles, these modified cells show other characteristics like mitochondria, endoplasmic reticulum, dictyosomes and glycogen granules. It should be stressed, however, that the typical differentiations of gland cells like glandular vesicles or of sensory cells like neurotubules and neurovesicles were not observed in these cells.

The apical ends of the modified cells are interconnected with neighbouring unmodified epidermis cells by well developed zonulae adhaerentes and more inconspicuous septate junctions. In these modified cells, only small amounts of the intracellular terminal web are present which anchor at the zonulae adhaerentes. The septate junctions are sometimes flanked by voluminous cisternae of the ER, a feature common to acoels (cf. TYLER 1984).

A slight constriction in form of a neck region of the modified cell is caused by the circular and longitudinal muscles of the body wall musculature (Figs. 2, 4 A).

In the more basal regions, the membranes of the cell and those of bordering cells are heavily interwoven (Figs. 3, 4 B). These adjacent cells are unmodified epidermal cells with their cell bodies, intra- or basiepidermal neurons and sensory cells, and, more interesting, many cytoplasmic processes of the central cellular digestive system, of the more peripheral located large "parenchymal" digestive cells, of subepidermal gland cells like rhabdoid gland cells as well as ramifications of the irregular subepidermal channel-like system of specialized glandular cells (see EHLERS 1992 a). In addition, developing or mature male gametes can also be situated near the basal regions of the modified epidermal cells.

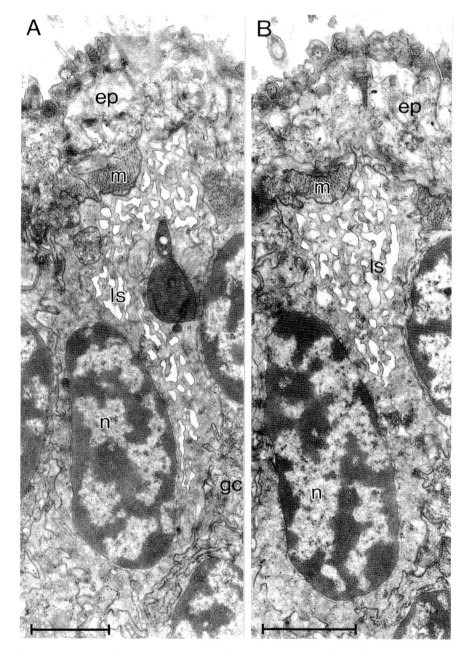

Fig. 3. Sections through the more proximal regions of a dermonephridium (the same as in Fig. 2) with its nucleus situated beneath the body wall musculature and with small isolated vesicles (in B) near the basal portions of the lacunar system. A. X 21 000; B. X 25 000.

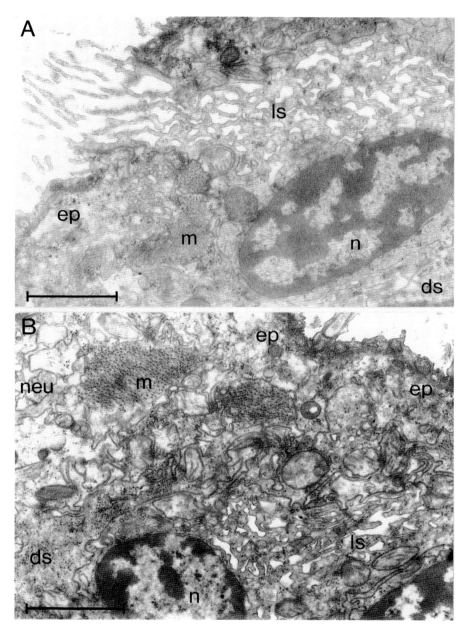

Fig. 4. Sections through two other dermonephridia with the peripheral parts of the lacunar system and the nucleus (in A) and with the basal parts of the lacunar system and the nucleus (in B). A. X 24 000; B. X 26 000.

D. Discussion

Up to now, the modified epidermal cells as described here have been unknown for any representative of the Plathelminthes.

It is most probable, that the occurrence of these cells is restricted to *Paratomella*, because extensive studies have been done on the fine structure of the epidermis and the non-epithelial cells of the body wall in all well-known taxa of the free-living Plathelminthes (see TYLER 1984; EHLERS 1985 and RIEGER et al. 1991 for references), including many species of the Acoela and of the Nemertodermatida, and no comparable differentiation has been found.

In the total absence of any cilia, these cells do not contribute to the ciliary locomotion of the animals. An adhesive function of the cells seems unlikely, because *Paratomella rubra* possesses well developed caudal haptocilia thoroughly described by TYLER (1973), see also EHLERS (1992 a), and because the apical thread-like projections of these cells are not strenghthened by filaments. Modified epidermal cells (anchor cells) in duo-gland adhesive systems of free-living Rhabditophora display fiber-cored microvilli (TYLER 1976, 1977, 1988; SOPOTT-EHLERS 1979; EHLERS 1985).

Such a statement is also corroborated by the fact, that any evidence for a glandular function of these cells is missing. Gland cells with well developed secretions exist in various body regions of *Paratomella* as described by CREZEE (1978) with the light microscope and by SMITH & TYLER (1986) and EHLERS (1992 a, b) from the EM-level. A sensory function of the cells is also unlikely, furthermore, these cells are not parts of the complicated cellular digestive system in *Paratomella* as described by SMITH & TYLER (1985) and EHLERS (1992 a). And I would like to stress, that these cells do not represent parts of the mouth opening or of gonopores nor are these cells injured resulting from a transfer of sperm from a partner in mating or from releasing zygotes (eggs) or from fission (paratomy).

The highly branching lacunar system and the apical thread-like cytoplasmic projections give reasons for the assumption of a nonspecialized (without further modification of the extracellular fluid) excretory function of these cells, which are called dermonephridia.

Cytoplasmic differentiations of a similar appearence are known from excretory organs or excretory cells in other Bilateria. A lacunar system occurs in the paranephrocytes of *Xenoprorhynchus*, a species of the Lecithoepitheliata (REISINGER 1968), in the tubule cells and in a nephroporus cell of protonephridia in gnathostomulids (LAMMERT 1985; according to this author the systems in the Gnathostomulida belong to the ER), in the terminal cells of several species of the Acanthocephala (e. g. DUNAGAN & MILLER 1986), in the labyrinth of the antennal glands of crustaceans (lit. in RIEGEL & COOK 1975) and in the so-called excretory organs of nematodes (e. g. NELSON et al. 1983). The lacunar system is also reminiscent of the spongiome system of ciliates (e. g. CHAPMAN & KERN 1983) and other protists (lit. in PATTERSON 1980). Microvilli-like projections exist in the nephroporus region of protonephridia in other Plathelminthes, e. g. in proseriates (EHLERS & SOPOTT-EHLERS 1987). In addition, the existence of small vesicles in the cytoplasm around the lacunar system, especially around its basal portions, corroborates this functional interpretation. At least several of these vesicles, which are interpreted as excretory vesicles, fuse with the lacunar system.

Nevertheless, there are fundamental differences between the dermonephridia in *Paratomella* and protonephridia in the Catenulida and the Rhabditophora, respectively.

In *Paratomella*, the cells are not arranged to a bilaterally symmetrical system like the protonephridia in the Rhabditophora (or many other Bilateria) or developed to an unpaired system comparable with the protonephridia in the Catenulida (see EHLERS 1985). A homologization of dermonephridia with any part of a protonephridium (terminal cell, tubule cell, nephroporus cell) in any species of the Bilateria is impossible. Furthermore, there is no evidence that the cells of *Paratomella* represent a precursor or a remnant of a protonephridial system as developed in other bilaterians.

There are strong arguments that excretory organs in form of protonephridia represent an evolutionary novelty of the taxon Bilateria (AX 1984, 1987). The protonephridia, which occur within the Bilateria, can be hypothesized as homologous organs (BARTOLOMAEUS & AX 1992). That means, the existence of such an excretory organ within the Plathelminthes is a plesiomorphic characteristic, inherited from the stem species of the Bilateria, whereas the absence of a protonephridium in the Nemertodermatida and the Acoela, which together constitute the taxon Acoelomorpha, is an autapomorphy of this taxon Acoelomorpha (EHLERS 1984, 1985).

Within the Acoelomorpha, this new type of an excretory system in form of modified epidermal cells with a lacunar system has been developed. Because the occurrence of the dermonephridia seems to be restricted to *Paratomella*, it is most probable that this unique feature is an autapomorphy of this taxon or of *P. rubra*, respectively (see EHLERS 1992 a).

As stated already, the Acoelomorpha lack any protonephridia. How can this absence be explained functionally?

Protonephridia as realized in other Bilateria display terminal cells with filter regions. In such a filter region, there exist fine molecular sieves in form of ECM barriers between a surrounding intercellular body cavity and the protonephridial lumen. That means, a protonephridium with a filter can only be realized in taxa which display extracellular matrices in intercellular spaces. The Acoela (and the Nemertodermatida in general as well) lack extracellular matrices in nearly all parts of the body (for lit. see RIEGER et al. 1991), in acoels (except *Paratomella*) fine matrices of a nonfibrillar appearance are only present in form of very thin capsules around the statocysts (EHLERS 1991, 1992 a, c). Thus, the absence of protonephridia may be correlated with the lack of extracellular matrices.

With the absence of an extracellular molecular sieve the modified epidermal cells of *Paratomella* are quite distinct from the ciliated terminal cells of protonephridia, where ultrafiltration occurs. In *Paratomella*, excretion seems to be an active process of secretion into the lacunar system of the modified epidermal cells, which are closely bordered to cells of the digestive system and to other cell types. There is no evidence of any further modification of extracellular fluids (e. g. reabsorption) in the more distal areas of the cells. However, the process of excretion may be influenced by the muscles (positive pressure) of the body wall musculature situated around the neck region of the modified epidermal cells.

Apparently, these cells of the marine *Paratomella* do not have a strong osmoregulatory function. This observation corresponds with findings in other plathelminth taxa like the Catenulida, where well developed protonephridia with an osmoregulatory function exist in limnic taxa (e. g. *Catenula, Stenostomum, Rhynchoscolex*), whereas marine taxa (e. g. *Retronectes*) have inconspicuous nephridia (EHLERS 1985). It would be interesting to study the fine structural organization of the very few acoels (e. g. *Oligochoerus limnophilus* Ax & Dörjes and *Limnoposthia polonica* (Kolasa & Faubel)) occurring in limnic environments in more detail. Perhaps discrete morphological features with an excretory (osmoregulatory) function can be found in limnic acoels.

Acknowledgements

I thank Mrs. K. Lotz for her valuable technical assistence and Mr. B. Baumgart for the realization of the drawing. Mrs. M. Olomski and Dr. J. Gottwald kindly provided the sediments taken at Gran Canaria in 1979 and 1980, respectively. Financial support was provided by the Akademie der Wissenschaften und der Literatur, Mainz.

Zusammenfassung

Paratomella rubra weist mehrere spezialisierte Epidermiszellen in verschiedenen Körperbereichen auf. Diesen nicht miteinander verbundenen Zellen, die als Dermonephridia bezeichnet werden, fehlen Cilien, das periphere Cytoplasma der Zellen ist in viele, schlanken Mikrovilli vergleichbare Fortsätze zergliedert. Große Bereiche der Zellen nimmt ein stark verzweigtes extrazelluläres Lakunensystem ein. Dieses ausgedehnte System besteht aus miteinander kommunizierenden Kanälchen und Vakuolen, umgeben von isolierten Vesikeln. Die Dermonephridia grenzen basal an Zellen des zentralen wie auch des peripheren „parenchymatischen" Verdauungssystems sowie an andere Zelltypen wie Drüsenzellen. Für die modifizierten Epidermiszellen wird eine exkretorische Funktion diskutiert.

Paratomella wie auch allen übrigen Vertretern der Acoelomorpha fehlen Protonephridien; dieser Mangel läßt sich funktionell erklären. Die Dermonephridia werden als eine Autapomorphie des Taxons *Paratomella* hypothetisiert.

Abbreviations

ci	locomotory cilium	gc	gland cell	neu	neurons
ds	digestive system	ls	lacunar system	sj	septate junction
e	epitheliosomes	m	muscle	tw	terminal web
ep	epidermis cell	n	nucleus	za	zonula adhaerens

References

Ax, P. (1984): Das phylogenetische System. Systematisierung der lebenden Natur aufgrund ihrer Phylogenese. G. Fischer, Stuttgart, New York, 349 p.
— (1987): The phylogenetic system. The systematization of organisms on the basis of their phylogenesis. John Wiley & Sons, Chichester, 340 p.
Bartolomaeus, T. & P. Ax (1992): Protonephridia and metanephridia – their relation within the Bilateria. Z. zool. Syst. Evolut.-forsch. 30, 21–45.
Chapman, G. B. & R. C. Kern (1983): Ultrastructural aspects of the somatic cortex and contractile vacuole of the ciliate, *Ichthyophthirius multifiliis* Fouquet. J. Protozool. 30, 481–490.
Crezee, M. (1978): *Paratomella rubra* Rieger and Ott, an amphiatlantic acoel turbellarian. Cah. Biol. Mar. 19, 1–9.
Dunagan, T. T. & D. M. Miller (1986): Ultrastructure of flame bulbs in male *Macracanthorhynchus hirudinaceus* (Acanthocephala). Proc. Helminthol. Soc. Wash. 53, 102–109.
Ehlers, B. & U. Ehlers (1980): Zur Systematik und geographischen Verbreitung interstitieller Turbellarien der Kanarischen Inseln. Mikrofauna Meeresboden 80, 1–23.
Ehlers, U. (1984): Phylogenetisches System der Plathelminthes. Verh. naturwiss. Ver. Hamburg (N. F.) 27, 291–294.
— (1985): Das phylogenetische System der Plathelminthes. G. Fischer, Stuttgart, New York, 317 p.
— (1989): The protonephridium of *Archimonotresis limophila* Meixner (Plathelminthes, Prolecithophora). Microfauna Marina 5, 261–275.
— (1991): Comparative morphology of statocysts in the Plathelminthes and the Xenoturbellida. Hydrobiologia 227, 263–271.
— (1992 a): On the fine structure of *Paratomella rubra* Rieger and Ott (Acoela) and the position of the taxon *Paratomella* Dörjes in a phylogenetic system of the Acoelomorpha (Plathelminthes). Microfauna Marina 7, 265–293.

- (1992 b): Frontal glandular and sensory structures in *Nemertoderma* (Nemertodermatida) and *Paratomella* (Acoela): ultrastructure and phylogenetic implications for the monophyly of the taxon Euplathelminthes (Plathelminthes). Zoomorphology (in press).
- (1992 c): Ultrastructure of the unencapsulated statocyst in *Paratomella rubra* Rieger and Ott (Plathelminthes, Acoela). Zoomorphology (in press).

EHLERS, U. & B. SOPOTT-EHLERS (1987): Zum Protonephridialsystem von *Invenusta paracnida* (Proseriata, Plathelminthes). Microfauna **3**, 377–390.

LAMMERT, V. (1985): The fine structure of protonephridia in Gnathostomulida and their comparison within Bilateria. Zoomorphology **105**, 308–316.

NELSON, F. K., P. S. ALBERT & D. L. RIDDLE (1983): Fine structure of the *Caenorhabditis elegans* secretory-excretory system. J. Ultrastruct. Res. **82**, 156–171.

PATTERSON, D. J. (1980): Contractile vacuoles and associated structures: their organization and function. Biol. Rev. **55**, 1–46.

REISINGER, E. (1968): *Xenoprorhynchus*, ein Modellfall für progressiven Funktionswechsel. Z. zool. Syst. Evolut.-forsch. **6**, 1–55.

RIEGEL, J. A. & M. A. COOK (1975): Recent studies of excretion in Crustacea. In: Excretion. Red.: A. WESSING. Fortschritte der Zoologie **23**, Heft 2/3, 48–75.

RIEGER, R. M. & J. A. OTT (1971): Gezeitenbedingte Wanderungen von Turbellarien und Nematoden eines nordadriatischen Sandstrandes. Vie Milieu, Suppl. **22**, 425–447.

RIEGER, R. M., S. TYLER, J. P. S. SMITH III & G. E. RIEGER (1991): Platyhelminthes: Turbellaria. In: Microscopic anatomy of invertebrates (Treatise ed.: F. W. HARRISON), Vol. **3** Platyhelminthes and Nemertinea. Eds.: F. W. HARRISON & B. J. BOGITSH. Wiley-Liss, New York, 7–140.

SMITH III, J. & S. TYLER (1985): The acoel turbellarians: kingpins of metazoan evolution or a specialized offshoot? In: The origins and relationships of lower invertebrates. Eds.: S. CONWAY MORRIS, J. D. GEORGE, R. GIBSON & H. M. PLATT. Oxford University Press, Oxford, 123–142.

SMITH III, J. P. S. & S. TYLER (1986): Frontal organs in the Acoelomorpha (Turbellaria): ultrastructure and phylogenetic significance. Hydrobiologia **132**, 71–78.

SOPOTT-EHLERS, B. (1979): Ultrastruktur der Haftapparate von *Nematoplana coelogynoporoides* (Turbellaria, Proseriata). Helgoländer wiss. Meeresunters. **32**, 365–373.

TYLER, S. (1973): An adhesive function for modified cilia in an interstitial turbellarian. Acta Zool. **54**, 139–151.
- (1976): Comparative ultrastructure of adhesive systems in the Turbellaria. Zoomorphologie **84**, 1–76.
- (1977): Ultrastructure and systematics: an example from turbellarian adhesive organs. Mikrofauna Meeresboden **61**, 271–286.
- (1984): Turbellarian plathelminths. In: Biology of the integument, Vol. 1 Invertebrates. Eds.: J. BEREITER-HAHN, A. G. MATOLTSY & K. S. RICHARDS. Springer, Berlin, 112–131.
- (1988): The role of function in determination of homology and convergence – examples from invertebrate adhesive organs. Fortschritte der Zoologie **36**, 331–347.

Prof. Dr. Ulrich Ehlers
II. Zoologisches Institut und Museum der Universität Göttingen
Berliner Straße 28, D-3400 Göttingen

On the fine structure of *Paratomella rubra* Rieger & Ott (Acoela) and the position of the taxon *Paratomella* Dörjes in a phylogenetic system of the Acoelomorpha (Plathelminthes)*

Ulrich Ehlers

Contents

Abstract	266
A. Introduction	266
B. Materials and Methods	267
C. Results	267
1. Epidermis and epidermal differentiations	267
2. Gland cells and glandular systems	273
3. Central and peripheral digestive systems	275
D. Discussion	279
1. Epidermis and epidermal cilia	279
2. Lack of a subepidermal matrix	282
3. Epidermal glands, frontal gland complex and subepidermal glandular system	283
4. Nervous system and sensory structures	284
5. Lack of protonephridia and existence of dermonephridia	286
6. Lack of a connective tissue	286
7. Digestive system	287
8. Vegetative and sexual reproduction	288
9. Phylogenetic conclusions: adelphotaxa-relationships of *Paratomella* and the Euacoela (= nov. monophylum) incl. *Hesiolicium*	289
Acknowledgements	291
Zusammenfassung	291
Abbreviations	291
References	292

---

* In memoriam Dr. Jürgen Dörjes († 7. 5. 1991), a foremost scholar in morphology and systematics of the Acoela

Abstract

The fine structural organizations of the epidermis, of epidermal receptors, of epidermal glands, of the subepidermal channel-like glandular system and of the digestive systems of the acoel *Paratomella rubra* are dealed with. These characteristics are discussed: epidermis and epidermal cilia; lack of a subepidermal matrix; epidermal glands; frontal gland complex; subepidermal glandular system; nervous system and sensory structures; lack of protonephridia and existence of dermonephridia; lack of a connective tissue; digestive system; vegetative reproduction (paratomy); several aspects of sexual reproduction.

A phylogenetic system of the Acoelomorpha is presented, in which *Paratomella* is the adelphotaxon of a monophylum named Euacoela and comprising all other acoels.

The monophyly of the taxa Acoelomorpha, Nemertodermatida, Acoela, *Paratomella* and Euacoela, respectively, is substantiated by several characteristics hypothesized as autapomorphies.

A. Introduction

The few species of the taxon *Paratomella* Dörjes, 1966 – *P. unichaeta* Dörjes, 1966, *P. rubra* Rieger & Ott, 1971 and *P. spec.* in Smith & Tyler 1986 – show several characteristics unknown or unusual for most other acoels.

From the light microscope vegetative (asexual) reproduction by paratomy and the existence of a complicated channel-like glandular system in the peripheral body regions have been described (Dörjes 1966; Rieger & Ott 1971; Crezee 1978). In addition, electron microscopy revealed several more peculiar features like caudal haptocilia (Tyler 1973), a cellular instead of a temporary or permanent syncytial central digestive system (Smith & Tyler 1985), a frontal glandular complex with three special types of gland cells instead of a frontal organ (Smith & Tyler 1986; Ehlers 1992 c), a statocyst without a capsule (Ehlers 1991, 1992 d), and the existence of dermonephridia (Ehlers 1992 a).

When discussing the systematic position of *Paratomella* on the basis of these and other known features, it becomes obvious, that the phylogeny of the Acoelomorpha in general has to be considered. Hence, a scheme of a phylogenetic system of the Acoelomorpha will be presented, in which all Acoela except *Paratomella* are united in a monophylum named Euacoela.

B. Material and Methods

All electron micrographs of *P. rubra* presented in this paper were obtained from specimens from the Canary Islands (for localities see EHLERS & EHLERS 1980; for EM-processing see EHLERS 1992 a). In addition, light microscopical observations on specimens of *P. rubra* from the Atlantic coasts of France and from the North Sea (localities in EHLERS & EHLERS 1980) were also taken into consideration. Moreover, I made use of all published original informations on *Paratomella unichaeta* (AX & SCHULZ 1959; DÖRJES 1966, 1968), on *Paratomella rubra* (RIEGER & OTT 1971; TYLER 1973, 1979, 1984; CREZEE & TYLER 1976; BOADEN 1977; CREZEE 1978; EHLERS & EHLERS 1980; SMITH & TYLER 1985; RIEGER et al. 1991; EHLERS 1992 a, c, d) and on *Paratomella spec.* (SMITH & TYLER 1986; RIEGER et al. 1991).

C. Results

This description of the fine structural organization of *P. rubra* concentrates on the epidermis and several epidermal differentiations, on the four different types of subepidermal gland cells and on the peripheral and central digestive systems. Other features like epidermal dermonephridia, the frontal glandular complex and the statocyst are described elsewhere (EHLERS 1992 a, c, d).

1. Epidermis and epidermal differentiations

The cellular and ciliated epidermis belongs to the so-called "insunk" type, i. e., most pericarya of the epidermal cells take a position below the peripheral body musculature (Figs. 1 B, 2, 6 A). But the nuclei of a few cells lie above the musculature (Fig. 2). Neither a lamina densa nor a lamina reticularis separate the epidermal cells from the underlying or neighbouring muscle cells, gland cells, nerve cells or other subepidermal cell types. All the epidermal cells are joined apically by well developed zonulae adhaerentes and by septate junctions. The intracellular terminal web anchor at the zonulae adhaerentes. More basally, the cell membranes of the epidermal cells most often interdigitate heavily.

The cilia of the epidermal cells show the organization typical of the Acoela. At the distal tips of the axonemes there exists a prominent shelf leading to a narrow diameter of the cilium in which the microtubule double numbers 4–7 no longer exist. The rootlet system of a cilium consists of a main rootlet pointing rostrally, a more indistinctive posteriorly pointing rootlet with fibre bundles running to the knee-like bends of neighbouring main rootlets, and of two lateral rootlets

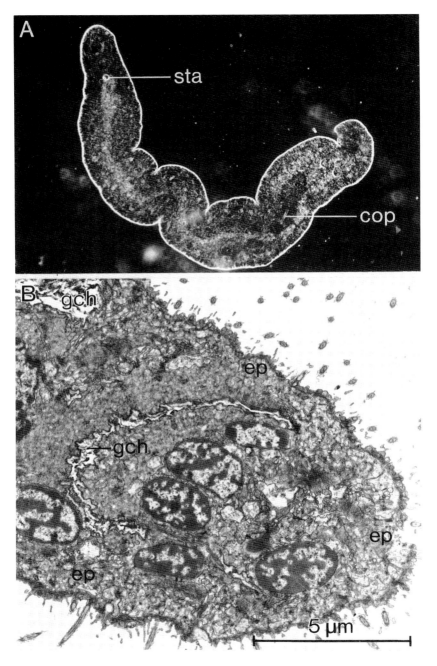

Fig. 1. *P. rubra.* A. Specimen in the beginning of male maturity (Phase-contrast micrograph). B. Low magnification micrograph of the epidermis with intraepidermal branches of the channel-like glandular system in oblique transverse section.

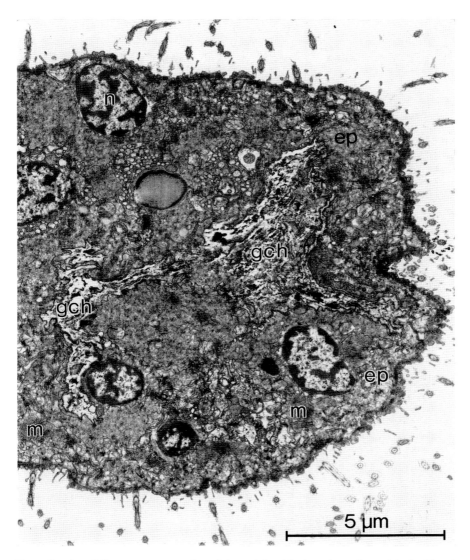

Fig. 2. *P. rubra*. Oblique transverse section of the peripheral body regions; the nuclei of epidermis cells take a position below (lower half) or above (upper half) the body wall musculature.

joining the tips of two posteriorly-laterally situated main rootlets (Fig. 3 C, D). Prominent accumulations of glycogens exist in the basal bodies of the cilia and in the basal portions of the main rootlets (Fig. 3 C–E).

Two other types of epidermal cells occur besides these typical ciliated epider-

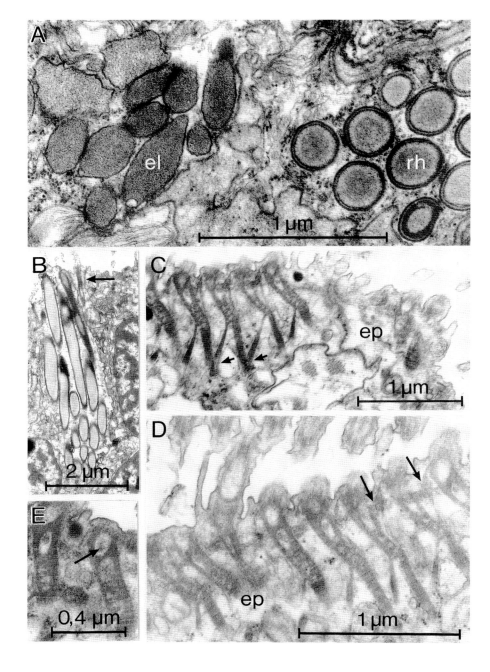

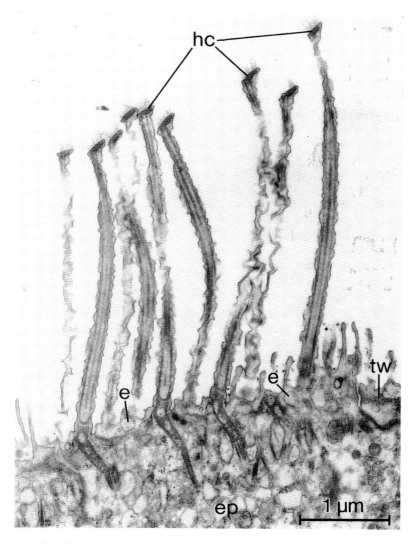

Fig. 4. *P. rubra.* Caudal adhesive plate with haptocilia in oblique longitudinal section and with epitheliosomes distal to the terminal web but without accumulations of dense rods.

◀ Fig. 3. *P. rubra.* A. Parts of two different epidermal glands with ellipsoids and with rhabdoids. B. Outleading neck of a rhabdoid gland cell, the arrow points to the rootlet of a monociliated noncollared epidermal receptor. C. Rootlet-system in oblique longitudinal section, the arrows point to contacts of lateral rootlets and neighbouring main rootlets. D. Oblique longitudinal section of the rootlet-system with fiber bundles originating from the caudal rootlet and running to the knee-like bends of adjacent main rootlets. E. Basal bodies and main rootlets with glycogen (arrow).

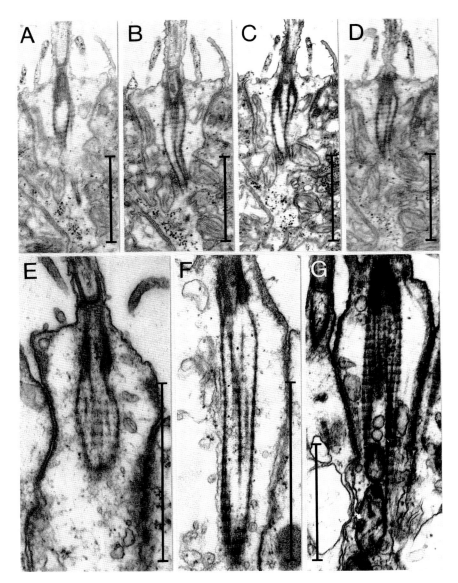

Fig. 5. A–D. *P. rubra*. Series of sections of a noncollared monociliated epidermal receptor with a specialized vertical rootlet. E–F. *Diopisthoporus psammophilus*. Sections of a receptor-type with a modified vertical rootlet identical with *P. rubra*. G. *Anaperus tvaerminnensis*. Epidermal receptor with a very similar, most probably identical vertical rootlet. Scale bars in A–G: 1 μm.

mal cells. First, unciliated epidermal cells with a lacunar system of a probable excretory function exist in various body regions (EHLERS 1992 a). Another type of modified cells can be found at the caudal adhesive plate. These cells bear haptocilia, specialized cilia for temporary adhesion. The haptocilia, which bend slowly in live material, are relatively straight and display mushroom-shaped distal ends, the flattened outside surfaces of the tips bear fibrous materials (Fig. 4). The basal portions of the haptocilia correspond to those of locomotory cilia, i. e., crook-like structural modifications of the axonemes are absent and the lateral rootlets of the rootlet systems are present. Furthermore, the terminal webs of the cells bearing haptocilia are not markedly different from those of other epidermal cells. Epitheliosomes are present in the most upper cytoplasm of the cells above the terminal webs (Figs. 4, 6 A). However, it should be noted that accumulations of small darkly staining rods were not observed, neither in unmodified epidermis cells nor in the cells of the adhesive plates.

The different types of epidermal receptors will not be discussed here in detail; observations on the statocyst and the nervous system are given by EHLERS (1992 d). By far the most common type of epidermal receptors are monociliary ones connected to the subepidermal nervous system, but bi- and tetraciliary receptors also exist (EHLERS 1992 c). All the receptors reach the body surface between the epidermal cells, penetrations of epidermal cells were not observed. The most common type of noncollared monociliary receptors show a specialized vertical rootlet composed of a peripheral sheet-like and slightly bended striated mantle and a central more straight and hollow element (Fig. 5 A–D).

2. Gland cells and glandular systems

DÖRJES (1966, p. 189/90) described four types of gland cells in *P. unichaeta*, these four types also occur in *P. rubra* and will be dealed with in the same order.

The first type of glands shows small rounded to rhabditoform and intensively staining secretion granules (Fig. 7 A) named target granules by SMITH & TYLER (1986, p. 76).

Gland cells of the second type show rounded to rhabditoform secretion granules of a slightly more greater diameter and of lesser staining secretions than the target granules (Fig. 3 A). SMITH & TYLER (1986) named these granules with homogeneous secretions ellipsoids.

The third type of glands, containing rod-shaped secretions (Figs. 3 A, B, 9), predominantly occurs in the lateral body portions but also in other regions like the frontal glandular complex. The secretion granules, which are termed rhabdoids by SMITH & TYLER (1986), resemble "true" rhabdites in having a structured cortex (Fig. 3 A) and a more homogenous medulla (but see SMITH & TYLER 1986,

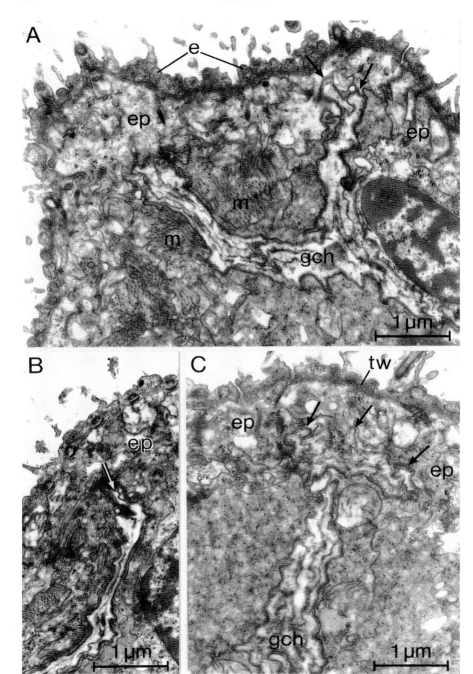

fig. 18). In the cortex region, there exists a moderate electron-dense fine outer layer just beneath the membranes of a granule, underlaid by a prominent membrane-like and more electron-dense lamella, inwards followed by an electron-light layer, most often of a faint lamellate appearance (Fig. 3 A).

The pericarya of all these three types of unicellular glands lie subepidermally, each gland neck projects to the epidermal surfaces between epidermal cells and releases the secretions to the exterior.

The fourth type of glandular cells is quite different. These cells do not have outleading necks and the secretions are of a non-granular nature (Figs. 1 B, 2, 6, 7, 8 A). The cell bodies, which take a position at the level of the other three gland types within the peripheral digestive tissues (see below), are scattered throughout the length of all specimens. The outlines of these remarkable cells are very irregular. Apparently, several of these cells are connected with each other thus forming a syncytium, but unicellular glands also occur. Altogether, these intensively staining glands form a channel-like and branching glandular system with many ramifications projecting to the epidermis. However, openings to the epidermal surface do not exist, the most distal areas of the branches terminate beneath the septate junctions of the adjacent epidermal cells (Fig. 6 A–C). Only few organelles like ER and mitochondria have been found in these glandular cells, mostly in the perinuclear regions (Fig. 7 A). The secretions are amorphous, aggregated into electron-dense flocculent masses or, more often, into electron-dense stratified deposits. Such an arrangement in strata can be seen in perinuclear regions (Fig. 8 A), but also in the intraepidermal branches (Fig. 6 A, C). The inner surfaces of the cell membranes are always covered with a thin uniform layer of these secretions. In short, the whole volume of such a cell or a syncytium is of a glandular nature, compartments with a less modified cytoplasm have not been found.

3. Central and peripheral digestive system

P. rubra does not show an epithelium-bounded digestive tract nor a syncytial digestive organization. The digestive system consists of two parts, a central and a peripheral one. Both systems are like "parenchymal" tissues, but can be distinguished from each other on the basis of their structural organizations.

The central digestive system consists of unicellular digestive cells with a highly fragmented cytoplasm. This cytoplasm forms many processes of varying diame-

---

◂ Fig. 6. *P. rubra.* A–C. Peripheral body regions with branches of the channel-like glandular system penetrating into the epidermis, the arrows point to the most peripheral terminations of these branches.

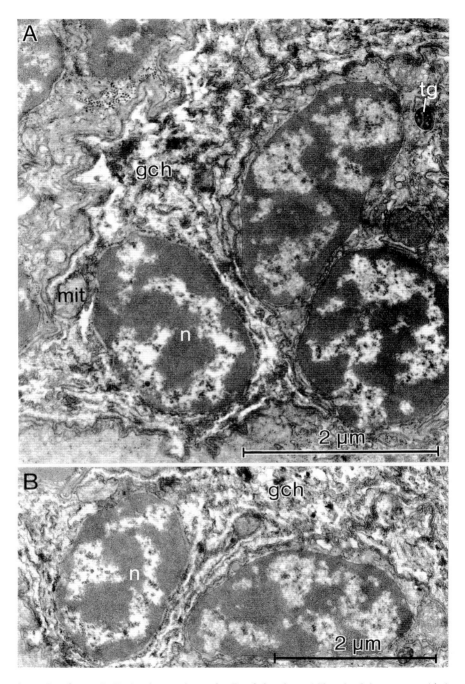

Fig. 7. *P. rubra*. A–B. Perinuclear regions of cells of the channel-like glandular system with its peculiar secretions. In A a target granule of a neighbouring target gland cell.

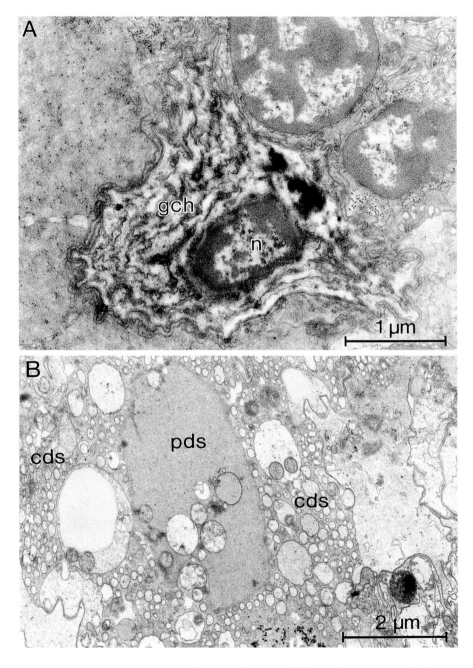

Fig. 8. *P. rubra*. A. Perinuclear region of a cell of the channel-like glandular system with electron-dense stratified secretions. B. Central parts of the digestive system with fragmented cytoplasmic areas of the central digestive cells and with a cell of the peripheral digestive system.

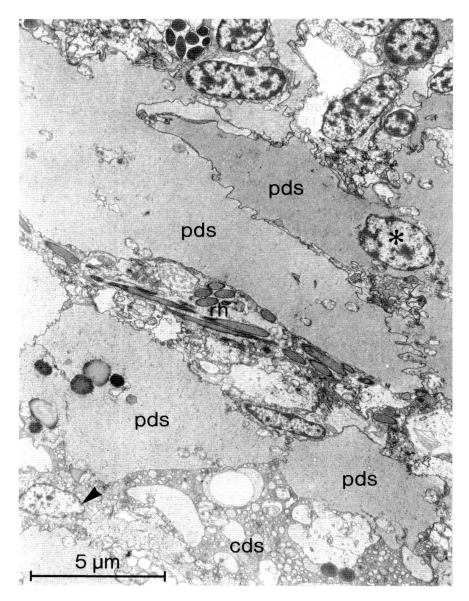

Fig. 9. *P. rubra.* Digestive system with the more voluminous cells of the peripheral digestive system (the asterisk marks a nucleus), in the lower half the fragmented cells of the central digestive system (the arrowhead points to a nucleus) and an epidermal rhabdoid gland cell penetrating the peripheral digestive system.

ters (Figs. 8 B, 9), the processes project into central extracellular "digestive" spaces filled with fluids of a moderate electron density. The larger ones of these processes contain organelles like mitochondria, but also lysosomal vesicles. The nuclei of these cells can be found within these regions (Fig. 8 B), each nucleus is surrounded by a thin layer of cytoplasm.

The cells of the peripheral digestive system are more voluminous (Figs. 8 B, 9), their cytoplasm is relatively homogeneous. Most of the organelles like ER and mitochondria are located in a perinuclear position. These cells, which also display lysosomal vacuoles, multivesicular bodies and lipid droplets, extend fine cytoplasmic projections into the central areas, i. e., the central body regions are filled with projections of both the central and the peripheral located digestive cells. In addition, the nuclear portions of at least several of these peripheral digestive cells also lie within this central body region (Fig. 9). And besides this, dorsoventral musculature and rhabdoid glands also occur here.

The peripheral digestive system extends peripherally to the body wall musculature, to the cell bodies of the epidermis cells and to the different gland cells. Stem cells and little differentiated digestive cells with small amounts of cytoplasm have been found in these peripheral subepidermal body regions.

P. rubra lacks any traces of an extracellular matrix in all body regions studied up to now. Furthermore, a peripheral "parenchyma" (without a digestive function) has not been found. Apparently, cells like "fixed parenchymal" cells or chordoid cells, both described as "parenchymal" cells, do not exist in *P. rubra*.

D. Discussion

The characteristics presented in chapter C and other known features of *P. rubra* will be discussed in comparison with the known characteristics of other acoels and nemertodermatids (see Table 1 and Fig. 10) as well as the Catenulida and Rhabditophora, respectively.

1. Epidermis and epidermal cilia

P. rubra has a multiciliary cellular epidermis, in which the locomotory cilia are not situated in pits of the cell surface. This situation is typical of the taxon Euplathelminthes (EHLERS 1985, 1986).

All the locomotory cilia show a tip with a pronounced shelf where the microtubule numbers 4–7 terminate, the remaining microtubules are capped by a flat plate. Such a characteristic is known for all representatives of the Acoela and for the Nemertodermatida, but not for any other taxon of the Plathelminthes

Table 1. Hypothesized autapomorphies of the monophyla named in Fig. 10 (p. 289)

Autapomorphies	Plesiomorphic alternatives
(1) Acoelomorpha	
(a) each epidermal cilium with an abrupt shelf-like narrowing near the distal tip (termination of microtubule doublets 4–7)	epidermal cilia without such a marked shelf near the distal tip
(b) rootlet system of an epidermal cilium consisting of a rostrally pointing main (bipartite?) rootlet with a knee-like bend and a posterior rootlet which splits into lateral fiber bundles that descend to and contact the main rootlets of two adjacent cilia	rootlet system of an epidermal cilium consisting of a straight rostral rootlet and of a straight caudal rootlet without any splitting
(c) each epidermal cilium with an accumulation of glycogen granules in its basal body and the adjacent parts of the main rootlet	epidermal cilia without prominent accumulations of glycogen granules
(d) far reaching absence of a basiepidermal basal lamina and a reticular lamina	existence of a fine, but distinct basiepidermal extracellular matrix
(e) existence of "pulsatile bodies" (= degenerating epidermal cells)	lack of "pulsatile bodies"
(f) lack of protonephridia	presence of protonephridia
(2) Nemertodermatida	
(a) existence of a statocyst with several parietal cells and two lithocytes (two statolithes)	lack of a statocyst
(b) monociliated spermatozoon with a prominent mitochondrial derivative	monociliated spermatozoa with unmodified mitochondria
(3) Acoela	
(a) rootlet system of an epidermal cilium with two lateral connections originating from the main rootlet	rootlet system of an epidermal cilium without lateral rootlets
(b) total lack of any basiepidermal extracellular matrix	existence of a discontinuous basiepidermal extracellular matrix
(c) existence of a statocyst with two parietal cells and one movable lithocyte (one statolith)	lack of a statocyst
(d) existence of a monociliated uncollared epidermal receptor-type with a specialized vertical rootlet system	absence of such an epidermal receptor-type
(e) digestive system without any digestive gland cells	intestinum with digestive gland cells
(f) spermatozoon "biciliated" (existence of two incorporated axonemes)	spermatozoon monociliated
(g) spermatozoon with dense ("acrosomal") granules	lack of dense granules in the spermatozoon

Table 1. (Continued)

(4) *Paratomella*	
(a) existence of a subepidermal channel-like glandular system	lack of such a glandular system
(b) existence of special types of epidermal glands (with target granules and rod-shaped rhabdoids), lack of mucoid glands	lack of these specialized epidermal glands but existence of mucoid glands
(c) existence of dermonephridia (described for *P. rubra*)	lack of specialized excretory differentiations
(d) existence of a caudal adhesive plate with specialized cilia (= haptocilia, described for *P. rubra*)	lack of a caudal adhesive plate and lack of haptocilia

(5) Euacoela	
(a) movable lithocyte of the statocyst with a marked ventral tubular system	lack of a pronounced tubular system in the lithocyte of the statocyst
(b) existence of a monociliated collared epidermal receptor-type with a swallow-nest like rootlet	absence of such an epidermal receptor-type
(c) at least central areas of the digestive system in form of a temporary or permanent digestive syncytium	cellular digestive system

(e. g. CREZEE & TYLER 1976; TYLER & RIEGER 1977; TYLER 1979; AX 1984; EHLERS 1985; RHODE et al. 1988). That means, this characteristic (no. 1 a in Table 1) can be hypothesized as an autapomorphy of the taxon Acoelomorpha.

The rootlets of the cilia of a cell are interconnected: each cilium has a main (bipartite?) rootlet with a pronounced knee-like bend which is joined by fiber bundles extending from posterior rootlets of neighbouring cilia. This situation is known from the Nemertodermatida and most acoels (e. g. TYLER & RIEGER 1977; AX 1984; EHLERS 1985), but not from other plathelminths, thus forming another autapomorphy (no. 1 b in Table 1) of the taxon Acoelomorpha. In addition, the rootlet system of a cilium in *P. rubra* shows two lateral rootlets which join the tips of the main rootlets in two adjacent cilia. This complicated rootlet system with lateral rootlets, which occurs in all acoels (e. g. HENDELBERG & HEDLUND 1974; TYLER 1979; AX 1984; EHLERS 1985; SMITH et al. 1986; RIEGER et al. 1991) but not in the Nemertodermatida, represents an autapomorphy (no. 3 a in Table 1) of the taxon Acoela. The basal bodies and the adjacent parts of the main rootlets show pronounced accumulations of glycogens in all nemertodermatids and acoels studied so far (EHLERS 1985); such accumulations are unknown from other plathelminths. Hence, this characteristic can be hypothesized as an autapomorphy (no. 1 c in Table 1) of the taxon Acoelomorpha.

P. rubra displays so-called "pulsatile bodies" (CREZEE 1978). These structures most probably represent degenerating epidermal cells with their cilia inside a vacuole. "Pulsatile bodies" also occur in *P. unichaeta* (DÖRJES 1966) and in many other acoels but also in nemertodermatids (e. g. DOREY 1965; SMITH et al. 1986; TYLER et al. 1989; EHLERS 1992 b), whereas these structures have not been observed in other plathelminths. The existence of "pulsatile bodies" is hypothesized as an autapomorphy of the taxon Acoelomorpha (no. 1 e in Table 1).

P. rubra possesses a caudal adhesive plate. Here, the cilia of the epidermal cells are modified to haptocilia (TYLER 1973) serving a temporary adhesive function. A caudal adhesive plate is also known from *P. unichaeta* and *P. spec.* (DÖRJES 1966; CREZEE & TYLER 1976) and a few other acoels (c. f. TYLER 1973, p. 148; CREZEE and Tyler 1976). Furthermore, haptocilia are described from *Hesiolicium inops* Crezee and Tyler, 1976, but the haptocilia of this species are grouped into pronounced papilla, the axonemes have a marked basal crook and the rootlet systems lack lateral rootlets. Thus, caudal adhesive plates and "haptocilia" of different organizations may have evolved several times independently within the Acoela, especially in those species inhabiting clean sediments. Possibly, the caudal adhesive plates in *P. unichaeta*, *P. rubra* and *P. spec.* are homologous features, thus constituting an autapomorphy (no. 4 d in Table 1) of the taxon *Paratomella*.

Membrane-bounded secretory granules (epitheliosomes) and a fibrillar terminal web exist in the apical cytoplasm of the epidermal cells in *P. rubra*. Comparable structures are well-known for many free-living Plathelminthes, including the Catenulida, and may represent differentiations already existent in the stem species of the Plathelminthes (EHLERS 1985). TYLER (1973, figs. 4, 6, 9 and 1974, fig. 7) described dark staining rods in the apical cytoplasm of the ventral cells in the caudal adhesive plate of *P. rubra* from North Carolina. Such accumulations of dense bodies have not been found in our material from different localities of the Canary Islands. Further studies must show, whether this inconspicuous feature is a constant difference of separated populations of *P. rubra* on both sides of the Atlantic.

2. Lack of a subepidermal matrix

The lack of a subepidermal basal lamina (lamina densa) and a lamina reticularis in *P. rubra* is typical of all acoels. The Nemertodermatida generally lack subepidermal extracellular matrices as well, but a very thin and discontinuous granular matrix has been found in several body regions of a few nemertodermatids (SMITH & TYLER 1985; RIEGER et al. 1991).

The absence of a typical ECM, especially of a fibrillar lamina reticularis, can be discussed from a phylogenetical and a functional point of view. Apparently, the stem species of the Plathelminthes possessed a weakly developed ECM (EHLERS 1985; the existence of an ECM with at least few collagenous materials is a basic feature for all the Metazoa). That means, the far-reaching absence of such a matrix in the Acoelomorpha and the total lack in the Acoela both represent apomorphic characteristics (no. 1 d and no. 3 b in Table 1).

But this absence can not be explained by the idea (RIEGER et al. 1991), that the Acoelomorpha (or the plathelminths in general) may have originated from acoelomate larvae or juveniles of a coelomate macrofaunal ancestor through progenesis: extracellular matrices are present in the larvae of all coelomate Bilateria investigated at the EM-level, moreover, subepidermal matrices already exist in the planula larvae of the Cnidaria.

It seems more plausible, that the loss of a fibrillar subepidermal ECM in the Acoelomorpha is correlated with the acquisition of a new apical strenghthened layer. Most probably, the interconnected rootlets of the epidermal cilia in the Acoelomorpha in connection with the cell web substitute the subepidermal lamina reticularis. In the Acoela, interconnecting rootlet-systems with lateral rootlets exist; these systems are not destroyed by vibrations of an ultrasonic cleaner (EHLERS 1985, table 22) and withstand resorption in degenerating epidermal cells for a long time (EHLERS 1992 b). That means, the total loss of any subepidermal matrices in the Acoela can be explained by the possession of this complex and strenghthened rootlet-system.

3. Epidermal glands, frontal gland complex and subepidermal glandular system

The three types of epidermal gland cells found in *P. rubra* are identical with those figured by SMITH & TYLER (1986, figs. 16–18) for *Paratomella spec.* Moreover, DÖRJES (1966) described three different types of epidermal glands in *P. unichaeta*, most probably the same three types as in *P. rubra* and *P. spec.* The glands of all the three types can be found throughout the length of the body in all the species of the taxon *Paratomella.* These glands open at the apical end of the specimens as well, forming a glandular frontal complex (SMITH & TYLER 1986; EHLERS 1992 c).

At least two of the three types of epidermal glands (target granular and rhabdoid gland cells) as described here are not known from any other acoels than the species of *Paratomella*, moreover, these glands are unknown for nemertodermatids as well. That means, the existence of glands with such special types of target granules and rod-shaped rhabdoids can be hypothesized as one autapomorphy (or as two autapomorphies) of the taxon *Paratomella* (no. 4 b in Table 1).

SMITH & TYLER (1986) regard the existence of a frontal gland complex (with a scattered arrangement of glands) in *Paratomella* as a derived phenomenon, in comparison to the existence of a frontal organ (with a collection of several mucus-secreting glands emerging through a common frontal pore) in most other acoels. SMITH & TYLER (l. c.) discuss the possibility that the situation in *Paratomella* is correlated with the process of paratomy. But this mode of vegetative reproduction seems to be a plesiomorphic characteristic of *Paratomella* or the plathelminths in general, respectively (see below). Most probably, the existence of a frontal gland complex is a basic feature of the taxon Euplathelminthes (see EHLERS 1992 c for a broader discussion), the lack (loss) of paratomy in the Acoela other than *Paratomella* may have facilitated the evolution of a multi-glandular frontal organ (a more concentrated collection of epidermal glands at the apical pole).

The subepidermal channel-like glandular system as described here (see also TYLER 1973, figs. 4 and 6) is a common feature for all species of *Paratomella*, but unknown for all other plathelminths. Without doubt, this feature is an outstanding autapomorphy of the taxon *Paratomella* (no. 4 a in Table 1).

At the light-microscopical level (cf. DÖRJES 1966) the intraepidermal ramifications of this irregular system of gland cells and glandular syncytia seem to open at the body surface. But this is not the case as can be seen at the EM-level: the most peripheral parts of the system terminate beneath the apical cell junctions of the epidermal cells. The lack of membrane-bounded glandular secretions within the cells or syncytia and the existence of the amorphous or flocculent secretions give the impressions of intercellular matrix compartments in plathelminths with voluminous intercellular body cavities. Perhaps, these systems may contribute to a strenghthening of the body in *Paratomella* as do special types of connective tissues in several other acoels (see below). But another function of the channel-like system seems also likely. BOADEN (1977) demonstrated the existence of heme in *Paratomella*. Most probably, this heme, which is responsible for the specimens reddish or red hue in life, is localized in this system of *P. rubra*.

4. Nervous system and sensory structures

The general organization of the nervous system in *P. rubra* is typical of many Acoela. The existence of a basi- or subepidermal plexus-like peripheral system besides a central one comprising an apical accumulation of neurones (a "brain") above the statocyst and several longitudinal nerves, which are interconnected cranially (DÖRJES 1966; CREZEE 1978), are plesiomorphic features for *P. rubra*.

In contrast to other acoels, the statocyst of *Paratomella rubra* lacks a capsule (EHLERS 1992 d). But there is no doubt, that the statocyst is of the "acoelan-type" (EHLERS 1985, 1991 – autapomorphy no. 3 c in Table 1). A discussion is given elsewhere (EHLERS 1992 d), whether the absence of a capsule in *Paratomella* is a derived phenomenon or whether this state represents the plesiomorphic situation in comparison with those acoels having an encapsulated statocyst. The statocysts in all acoels except *Paratomella* display a central lithocyte with a pronounced ventral tubular system (FERRERO 1973; EHLERS 1985, 1991, 1992 d). Such a tubular system is unknown from any other type of statocysts e. g. in the Catenulida or the Nemertodermatida. By out-group comparison, the lack of a tubular system in the statocyst of *Paratomella* is a plesiomorphy whereas the existence of this complex differentiation can be hypothesized as an autapomorphy (no. 5 a in Table 1) of a taxon Euacoela comprising all other acoels.

Mostly monociliated, but also bi- and tetraciliated epidermal receptors have been found in *Paratomella rubra* and *Paratomella spec.* (SMITH & TYLER 1986; EHLERS 1992 c). The occurrence of monociliated receptors is widespread within all taxa of the Acoelomorpha and other plathelminths and represents a plesiomorphous characteristic of *Paratomella*. Bi- and tetraciliated receptors are also known from the nemertodermatids (RIEGER et al. 1991), whereas acoels other than *Paratomella* lack tetraciliated and – with the exception of a very few species – also biciliated epidermal receptors.

Two alternative hypotheses can be made on the basis of this known distribution of bi- and tetraciliated receptor-types within the Acoelomorpha.

a) Such receptors have been present in the stem species of the taxon Acoelomorpha and have been passed on to the Nemertodermatida and *Paratomella*, respectively. Then the lack of tetra- (and bi-) ciliated receptors in other acoels would be an apomorphic condition.

b) Only monociliated epidermal receptors have been present in the stem species of the taxon Acoelomorpha and have been passed on to all subordinated taxa. Then multiciliated receptors in *Nemertoderma* and *Paratomella* would have been evolved convergently.

Without further researches no statement can be made whether the first or the second hypothesis is of a higher probability.

In *Paratomella*, mono-, bi- and tetraciliated receptors do not penetrate epidermal cells but reach the surface of the specimens between epidermal cells, without doubt a plesiomorphic characteristic (SOPOTT-EHLERS 1984; EHLERS 1985).

The most common type of noncollared monociliary receptors of *P. rubra* as described in this paper is also known from several other acoels (Fig. 5 E– G; see also BEDINI et al. 1973; EHLERS 1985, 1986; POPOVA & MAMKAEV 1987; RAIKOVA 1989; RIEGER et al. 1991, fig. 15 B), but unknown from the nemertodermatids and other plathelminths. The most plausible hypothesis is, that such a type of monociliated receptors represent an evolutionary novelty of the taxon Acoela (no. 3 d in Table 1). Within the Acoela excl. *Paratomella*, additional epidermal receptors occur, for example collared monociliary receptors with a "swallow's nest" structure below the basal body (BEDINI et al. 1973; EHLERS 1985, 1986; POPOVA & MAMKAEV 1987; RAIKOVA 1989). The existence of such a type of a collared receptor with a complex rootlet may represent an autapomorphy (no. 5 b in Table 1; see also EHLERS 1985) of a taxon comprising all Acoela except *Paratomella*.

5. Lack of protonephridia and existence of dermonephridia

All acoelomorphs lack protonephridia, whereas such nephridial organs are present in the Catenulida and the Rhabditophora, respectively. By out-group comparison, the absence of protonephridia can be hypothesized as an autapomorphy (no. 1 f in Table 1) of the taxon Acoelomorpha (see AX 1984; EHLERS 1985 and 1992a for more detailed discussions).

No other specialized differentiations with an excretory function are known for the Nemertodermatida and the Acoela with the exception of *Paratomella*. In *P. rubra*, dermonephridia (specialized epidermal cells of a probable excretory function) have been found (EHLERS 1992 a). The existence of these conspicuous cells is hypothesized as an autapomorphy (no. 4 c in Table 1) of *P. rubra* or of the taxon *Paratomella* (if these cells occur in *P. unichaeta* as well).

6. Lack of a connective tissue

P. rubra lacks any connective ("parenchymal") tissues like most other representatives of the Acoelomorpha.

The absence of any kind of special parenchymal cells is a plesiomorphous feature, inherited from the stem species of the taxon Plathelminthes (EHLERS 1985), whereas the lack of any traces of extracellular matrices, especially of collagens, around muscle cells, nerve cells and other subepidermal cell types as well as the "insunk" cell bodies of epidermal gland cells can be hypothesized as an autapomorphy (no. 3 b in Table 1) of the taxon Acoela (see above).

Probably, the channel-like glandular system in *Paratomella* corresponds functionally with the large, highly branched "fixed parenchymal cells" (for lit. see RIEGER 1985 and RIEGER et al. 1991) or the vacuolated and also branched "chordoid cells" known from several other acoels. All these "parenchymal" cell types are unique derived features, their existence can not be hypothesized as a basic characteristic of the Acoela or the Acoelomorpha, respectively.

7. Digestive system

A simple mouth opening as in *Paratomella* can be found in many other acoels as well, but also in nemertodermatids. Most probably, such an entrance into the digestive system represents a plesiomorphic characteristic for the Acoelomorpha or the Plathelminthes in general, respectively (DOE 1981; RIEGER 1981; EHLERS 1985).

A cellular gut of an epithelial nature and with a defined central digestive cavity occurs in several representatives of the Nemertodermatida, e. g. in *Nemertinoides* (RISER 1987) and *Flagellophora* (RIEGER et al. 1991, fig. 22 B), whereas other nemertodermatids display a less epithelial gut without a pronounced lumen (EHLERS 1985). Within the Acoela, *Paratomella* is the only taxon for which a complete cellular, but not epithelial digestive system has been described (see also SMITH & TYLER 1985; RIEGER et al. 1991).

In contrast to the Nemertodermatida (and the Catenulida and Rhabditophora as well), *Paratomella* and all other representatives of the Acoela lack any digestive gland cells. This derived characteristic is hypothesized as an autapomorphy (no. 3 e in Table 1) of the taxon Acoela.

Furthermore, *Paratomella* lacks any intestinal ciliation. Ciliated guts are present in the Catenulida and the Rhabditophora, respectively. For the Nemertodermatida, gastrodermal cilia have not been demonstrated unambiguously (cf. RIEGER et al. 1991, p. 72) and the Acoela lack any intestinal ciliation definitely. That means, the stem species of the Acoelomorpha may have had an unciliated digestive tract already, an apomorphic characteristic for this taxon. But this possible autapomorphy is not included in Table 1 because of the uncertainty in the nemertodermatids.

SMITH & TYLER (1985) discussed the different organizations of the digestive systems within the Acoelomorpha in great detail. There are clear evidences that the digestive systems in all acoels except *Paratomella* show a syncytial nature, at least in the more central areas. Such digestive systems may be a temporary structure formed only after ingestion of food, or may be a permanently present differentiation (see also MAMKAEV & MARKOSA 1979; MARKOSA 1987; RAIKOVA 1987 a, b). The syncytia may be shed when digestion has finished leaving a large central cavity, but RAIKOVA (l. c.) observed a permanent central cavity surrounded by temporary syncytia in *Actinoposthia*. In any case, the existence of a temporary or permanent syncytial organization of at least parts of the digestive system is a derived feature, which can be hypothesized as an evolutionary novelty (autapomorphy no. 5 c in Table 1) of a taxon Euacoela comprising all acoels except *Paratomella*.

In *P. rubra*, the cellular digestive system consists of a central and a peripheral one. Apparently, the cells of the central one, which show a cytoplasm fragmented into many small areas, are mainly involved in digestion (see also SMITH & TYLER 1985). But cytoplasmic extensions of the more peripheral cells also project towards the central digestive lumen. These peripheral cells, which display stored nutrients (lipids) and multivesicular bodies as well, can be found in the central body region as well.

Nevertheless, the peripheral cells differ from the central ones by their sizes and their "parenchymal" appearances. Two different kinds of digestive tissues (often called central syncytium and peripheral wrapping cells) are known from many acoels. Such an organization may represent a basic feature of the Acoela.

8. Vegetative and sexual reproduction

Paratomy is typical of the species of the taxon *Paratomella*, but unknown for other acoels. Most probably, this mode of vegetative (asexual) reproduction is a plesiomorphic characteristic of *Paratomella*: paratomy occurs within the Catenulida and the Rhabditophora as well (EHLERS 1985; RIEGER 1986).

With respect to the sexual reproduction, *Paratomella* displays several plesiomorphic features as well, inherited from the stem species of the Plathelminthes. Apparently, the hermaphroditic species do not have gonads, but male and female germ layers without special tunica or follicle cells. But mixed layers are lacking, both layers are separated from each other (own observations; see also DÖRJES 1966; CREZEE 1978). The lack of any accessory female organs is another plesiomorphic characteristic of the taxon *Paratomella*.

Own preliminary observations on the ultrastructure of the male gametes show clearly, that these cells are of the well-known acoelan type (HENDELBERG 1977). Each mature sperm cell of *P. rubra* displays two axonemes of the 9X2 + 2 type (two "incorporated cilia") within its cytoplasm. And the cells have electron-dense granules ("acrosomal" granules) as do have other acoels. Both features are unknown from the Nemertodermatida and are apomorphic in comparison with the Nemertodermatida (see AX 1984 and EHLERS 1985 for further discussions), thus constituting two autapomorphies (no. 3 f and no. 3 g in Table 1) of the taxon Acoela.

Only one more point concerning the reproduction will be mentioned here. It is well-known, that the eggs of all acoels display a characteristic duett cleavage (lit. in EHLERS 1985, p. 146). Unfortunately, nothing is known about the cleavages of the Nemertodermatida and of *Paratomella*. It is therefore not possible to state, whether this apomorphic cleavage constitutes an autapomorphy of the taxon Acoelomorpha or of the taxon Acoela or only of the taxon Euacoela (see below), respectively.

9. Phylogenetic conclusions: adelphotaxa-relationships of *Paratomella* and the Euacoela (= nov. monophylum) incl. *Hesiolicium*

The systematic position of *Paratomella* in a phylogenetic system of the Plathelminthes can be hypothesized on the basis of all known characteristics discussed above.

There is no doubt that the taxon *Paratomella* is a member of the monophylum Acoela. Such a relationship is corroborated by several characteristics, for example the existence of lateral rootlets in the rootlet systems of epidermal cilia, the organization of the statocyst, the existence of specialized epidermal receptors or the occurrence of spermatozoa with two incorporated axonemes (see Table 1).

Within the Acoela, *Paratomella* displays characteristics which represent plesiomorphies in comparison with all other acoels. *Paratomella* is the only known acoelan taxon with an entirely cellular digestive system. All other acoels display

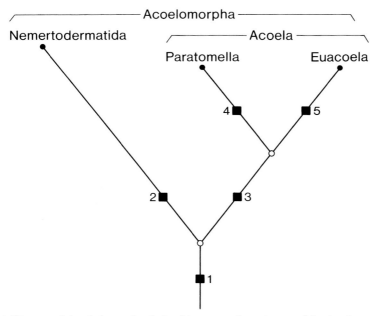

Fig. 10. Diagram of the phylogenetic relationships among the main taxa of the Acoelomorpha. Black squares 1–5 are hypothesized autapomorphies listed in Table 1 (p. 280).

small or larger syncytial areas within their digestive systems (cf. also RIEGER et al. 1991). The existence of a syncytium can be explained as one autapomorhy of a taxon for which the name Euacoela is introduced (for other hypothesized autapomorphies of the Euacoela see Table 1). This monophylum Euacoela is the adelphotaxon of the taxon *Paratomella*.

The hypothesis of such a position of *Paratomella* within the Acoelomorpha is not a new one. SMITH & TYLER (1985, fig. 8.6) proposed a transformation series for the digestive systems in the Acoelomorpha in which *Paratomella* takes an intermediate position between the Nemertodermatida and all other acoels. And RIEGER (1986, p. 40) stated, that *Paratomella* "appears to be most similar to the stem species" (of the Acoela) "on the basis of characters of the digestive parenchyma".

DÖRJES (1966) erected the family Paratomellidae for the new genus *Paratomella*. CREZEE & TYLER (1976) included *Hesiolicium* into the taxon Paratomellidae on the basis of the existence of haptocilia. As discussed above, the "haptocilia" in *Hesiolicium* and *Paratomella* most probably are not homologous structures. Other characteristics (absence of a pharynx, absence of female accessory organs) common to *Paratomella* and *Hesiolicium* are symplesiomorphies (see above). On the other hand, both taxa differ in many features (see diagnoses by CREZEE & TYLER l. c.) and *Hesiolicium* lacks the characteristic autapomorphies (channel-like glandular system, special types of epidermal glands) of *Paratomella*. The existence of a central digestive syncytium in *Hesiolicium* (see also SMITH & TYLER 1985) demonstrates, that this taxon belongs to the Euacoela.

The taxon Paratomellidae becomes monotypic when *Hesiolicium* is excluded. Then, the taxon Paratomellidae includes the single subtaxon *Paratomella*. But giving the same monophyletic entity more than one name is redundant. Hence, the name Paratomellidae is not used in the phylogenetic system of the Acoelomorpha as presented in Fig. 10.

The monophyly of the taxon Acoelomorpha can be hypothesized on the basis of several autapomorphies (see Table 1). ROHDE (1990) added the taxon *Xenoturbella* Westblad ("Xenoturbellida") to the Acoelomorpha, mainly because *Xenoturbella* displays epidermal cilia with distal shelfs corresponding with the Acoelomorpha and because of assumed similarities of the epidermal rootlet systems. But, as pointed out by PEDERSEN & PEDERSEN (1988) "the ciliary rootlet systems" (of the Nemertodermatida and *Xenoturbella*) "are quite different from each other". And there exist several well-known features (listed in EHLERS 1991) which clearly show that *Xenoturbella* can not be placed within the Acoelomorpha or the Plathelminthes at all. Thus, the shelfs in the axonemes of epidermal cilia in the Acoelomorpha and *Xenoturbella*, respectively, must be hypothesized as a non-homologous feature (sensu AX 1987, p. 159).

Acknowledgements

I thank Mrs. K. Lotz and Mrs. E. Hildenhagen-Brüggemann for their valuable technical assistences and Mr. B. Baumgart for his help with the figures. Mrs. M. Olomski and Dr. J. Gottwald kindly provided sediments with specimens of *P. rubra* from Gran Canaria. Financial support was provided by the Akademie der Wissenschaften und der Literatur, Mainz.

Zusammenfassung

Für das Acoel *Paratomella rubra* wird die ultrastrukturelle Organisation der Epidermis, der epidermalen Rezeptoren, der epidermalen Drüsen, des subepidermalen Drüsensystems und des Verdauungssystems dargestellt. Diskutiert werden folgende Merkmale: die Epidermis und die epidermalen Cilien; der Mangel einer subepidermalen Matrix; die epidermalen Drüsen; der Frontaldrüsenkomplex; das subepidermale Drüsensystem; das Nervensystem und die sensorischen Strukturen; der Mangel von Protonephridien und die Existenz von Dermonephridien; der Mangel eines Bindegewebes; das Verdauungssystem; die vegetative Vermehrung (Paratomie) und einige Aspekte der sexuellen Fortpflanzung.

Es wird ein phylogenetisches System der Acoelomorpha dargestellt, in dem *Paratomella* das Adelphotaxon eines Monophylums mit dem Namen Euacoela bildet; den Euacoela gehören alle übrigen Acoela an.

Die Monophylie der Taxa Acoelomorpha, Nemertodermatida, Acoela, *Paratomella* und Euacoela läßt sich jeweils über mehrere hypothetisierte Autapomorphien begründen.

Abbreviations

cds	central cellular digestive system	m	muscle
cop	male copulatory organ	mit	mitochondrion
e	epitheliosome(s)	n	nucleus
el	ellipsoids and ellipsoid gland cell	pds	peripheral cellular digestive system
ep	epidermis cell(s)	rh	rhabdoids and rhabdoid gland cell
gch	gland cell(s) of the channel-like glandular system	sta	statocyst
		tg	target granule(s)
hc	haptocilia	tw	terminal web

References

Ax, P. (1984): Das phylogenetische System. Systematisierung der lebenden Natur aufgrund ihrer Phylogenese. G. Fischer, Stuttgart, New York, 349 p.
- (1987): The phylogenetic system. The systematization of organisms on the basis of their phylogenesis. John Wiley & Sons, Chichester, 340 p.

Ax, P. & E. Schulz (1959): Ungeschlechtliche Fortpflanzung durch Paratomie bei acoelen Turbellarien. Biol. Zentralbl. 78, 613–622.

Bedini, C., E. Ferrero & A. Lanfranchi (1973): The ultrastructure of ciliary sensory cells in two Turbellaria Acoela. Tissue Cell 5, 359–372.

Boaden, P. J. S. (1977): Thiobiotic facts and fancies. (Aspects of the distribution and evolution of anaerobic meiofauna). Mikrofauna Meeresboden 61, 45–63.

Crezee, M. (1978): *Paratomella rubra* Rieger and Ott, an amphiatlantic acoel turbellarian. Cah. Biol. Mar. 19, 1–9.

Crezee, M. & S. Tyler (1976): *Hesiolicium* gen. n. (Turbellaria Acoela) and observations on its ultrastructure. Zool. Scr. 5, 207–216.

Doe, D. A. (1981): Comparative ultrastructure of the pharynx simplex in Turbellaria. Zoomorphology 97, 133–193.

Dörjes, J. (1966): *Paratomella unichaeta* nov. gen. nov. spec., Vertreter einer neuen Familie der Turbellaria Acoela mit asexueller Fortpflanzung durch Paratomie. Veröffentl. Inst. Meeresforsch. Bremerhaven Sonderband 2, 187–200.
- (1968): Die Acoela (Turbellaria) der deutschen Nordseeküste und ein neues System der Ordnung. Z. zool. Syst. Evolut.-forsch. 6, 56–452.

Dorey, A. E. (1965): The organization and replacement of the epidermis in acoelous turbellarians. Quart. J. micr. Sci. 106, 147–172.

Ehlers, B. & U. Ehlers (1980): Zur Systematik und geographischen Verbreitung interstitieller Turbellarien der Kanarischen Inseln. Mikrofauna Meeresboden 80, 1–23.

Ehlers, U. (1985): Das phylogenetische System der Plathelminthes. G. Fischer, Stuttgart, New York, 317 p.
- (1986): Comments on a phylogenetic system of the Platyhelminthes. Hydrobiologia 132, 1–12.
- (1991): Comparative morphology of statocysts in the Plathelminthes and the Xenoturbellida. Hydrobiologia 227, 263–271.
- (1992 a): Dermonephridia – modified epidermal cells with a probable excretory function in *Paratomella rubra* (Acoela, Plathelminthes). Microfauna Marina 7, 253–264.
- (1992 b): "Pulsatile bodies" in *Anaperus tvaerminnensis* (Luther, 1912) (Acoela, Plathelminthes) are degenerating epidermal cells. Microfauna Marina 7, 295–310.
- (1992 c): Frontal glandular and sensory structures in *Nemertoderma* (Nemertodermatida) and *Paratomella* (Acoela): ultrastructure and phylogenetic implications for the monophyly of the taxon Euplathelminthes (Plathelminthes). Zoomorphology (in press).
- (1992 d): Ultrastructure of the unencapsulated statocyst in *Paratomella rubra* Rieger and Ott (Plathelminthes, Acoela). Zoomorphology (in press).

Ferrero, E. (1973): A fine structural analysis of the statocyst in Turbellaria Acoela. Zool. Scr. 2, 5–16.

Hendelberg, J. (1977): Comparative morphology of turbellarian spermatozoa studied by electron microscopy. Acta Zool. Fennica 154, 149–162.

Hendelberg, J. & K.-O. Hedlund (1974): On the morphology of the epidermal ciliary rootlet system of the acoelous turbellarian *Childia groenlandica*. Zoon 12, 13–24.

Mamkaev, Yu. V. & T. G. Markosova (1979): Electron microscopic studies of the parenchyma in some representatives of the Acoela. Proc. Zool. Inst. Acad. Sci. USSR 84, 7–12 (In Russian).

Markosova, T. G. (1987): Pathes of intracellular digestion and transport in the turbellarian *Oxyposthia praedator* Ivanov. Tr. Zool. Inst. Akad. Nauk SSSR 167, 79–84 (In Russian).

Popova, N. V. & Yu. V. Mamkaev (1987): On types of sensillae in the acoel turbellarians. Tr. Zool. Inst. Akad. Nauk SSSR 167, 85–89 (In Russian).

PEDERSEN, K. J. & L. R. PEDERSEN (1988): Ultrastructural observations on the epidermis of *Xenoturbella bocki* Westblad, 1949; with a discussion of epidermal cytoplasmic filament systems of invertebrates. Acta Zool. (Stockh.) **69**, 231–246.
RAIKOVA, O. I. (1987 a): Ultrastructural organisation of the digestive system of the acoel turbellarian *Actinoposthia beklemischevi* Mamkaev. Tr. Zool. Inst. Akad. Nauk SSSR **167**, 72–78 (In Russian).
– (1987 b): Ultrastructure of the digestive parenchyma in *Actinoposthia beklemischevi* (Turbellaria, Acoela). Dokl. Acad. Nauk SSSR **293**, 1503–1505.
– (1989): Ultrastructure of the nervous system and sensory receptors of acoel turbellarians. Tr. Zool. Inst. Akad. Nauk SSSR **195**, 36–46 (In Russian).
RIEGER, R. M. (1981): Morphology of the Turbellaria at the ultrastructural level. Hydrobiologia **84**, 213–229.
– (1985): The phylogenetic status of the acoelomate organization within the Bilateria: a histological perspective. In: The origins and relationships of lower invertebrates. Eds.: S. CONWAY MORRIS, J. D. GEORGE, R. GIBSON & H. M. PLATT. Oxford University Press, Oxford, 101–122.
– (1986): Asexual reproduction and the turbellarian archetype. Hydrobiologia **132**, 35–45.
RIEGER, R. M. & J. A. OTT (1971): Gezeitenbedingte Wanderungen von Turbellarien und Nematoden eines nordadriatischen Sandstrandes. Vie Milieu, Suppl. **22**, 425–447.
RIEGER, R. M., S. TYLER, J. P. S. SMITH III & G. E. RIEGER (1991): Platyhelminthes: Turbellaria. In: Microscopic anatomy of invertebrates (Treatise ed.: F. W. HARRISON), Vol. 3 Platyhelminthes and Nemertinea. Eds.: F. W. HARRISON & B. J. BOGITSH. Wiley-Liss, New York, 7–140.
RISER, N. W. (1987): *Nemertinoides elongatus* gen. n., sp. n. (Turbellaria: Nemertodermatida) from coarse sand beaches of the western North Atlantic. Proc. Helminthol. Soc. Wash. **54**, 60–67.
ROHDE, K. (1990): Phylogeny of Platyhelminthes, with special reference to parasitic groups. Int. J. Parasitol. **20**, 979–1007.
ROHDE, K., N. WATSON & L. R. G. CANNON (1988): Ultrastructure of epidermal cilia of *Pseudactinoposthia* sp. (Platyhelminthes, Acoela): implications for the phylogenetic status of the Xenoturbellida and Acoelomorpha. J. Submicrosc. Cytol. Pathol. **20**, 759–767.
SMITH III, J. & S. TYLER (1985): The acoel turbellarians: kingpins of metazoan evolution or a specialized offshoot? In: The origins and relationships of lower invertebrates. Eds.: S. CONWAY MORRIS, J. D. GEORGE, R. GIBSON & H. M. PLATT. Oxford University Press, Oxford, 123–142.
SMITH III, J. P. S. & S. TYLER (1986): Frontal organs in the Acoelomorpha (Turbellaria): ultrastructure and phylogenetic significance. Hydrobiologia **132**, 71–78.
SMITH III, J. P. S., S. TYLER & R. RIEGER (1986): Is the Turbellaria polyphyletic? Hydrobiologia **132**, 13–21.
SOPOTT-EHLERS, B. (1984): Epidermale Collar-Receptoren der Nematoplanidae und Polystyliphoridae (Plathelminthes, Unguiphora). Zoomorphology **104**, 226–230.
TYLER, S. (1973): An adhesive function for modified cilia in an interstitial turbellarian. Acta Zool. **54**, 139–151.
– (1979): Distinctive features of cilia in metazoans and their significance for systematics. Tissue Cell **11**, 385–400.
– (1984): Turbellarian platyhelminths. In: Biology of the integument, Vol. 1 Invertebrates. Eds.: J. BEREITER-HAHN, A. G. MATOLTSY & K. S. RICHARDS. Springer, Berlin, 112–131.
TYLER, S., J. K. GRIMM & J. P. S. SMITH III (1989): Dynamics of epidermal wound repair in acoel turbellarians – the role of pulsatile bodies. Am. Zool. **29**, 115 A.
TYLER, S. & R. M. RIEGER (1977): Ultrastructural evidence for the systematic position of the Nemertodermatida (Turbellaria). Acta Zool. Fennica **154**, 193–207.

Prof. Dr. Ulrich Ehlers
II. Zoologisches Institut und Museum der Universität Göttingen
Berliner Straße 28, D-3400 Göttingen

"Pulsatile bodies" in *Anaperus tvaerminnensis* (Luther, 1912) (Acoela, Plathelminthes) are degenerating epidermal cells

Ulrich Ehlers

Contents

Abstract	295
A. Introduction	296
B. Materials and Methods	299
C. Results	299
D. Discussion	306
Acknowledgements	308
Zusammenfassung	308
Abbreviations	308
References	309

Abstract

Wild specimens of *Anaperus tvaerminnensis* (Luther, 1912) display so-called "pulsatile bodies", which are located within the epidermis and within subepidermal digestive tissues. These bodies are not epidermal replacement cells, but are degenerating epidermal cells, which are withdrawn from the epidermis and which become enclosed in vacuoles of the central syncytial digestive system for resorption.

Peripherally situated cilia of a degenerating epidermal cell become included and are resorbed in the cytoplasm of such a body. However, most of the ciliary axonemes remain intact and become enclosed in one large vacuole. Cilia of such cells being withdrawn into subepidermal tissues are no longer capable for pulsation: the membrane of the vacuole surrounding the axonemes becomes fragmented, all the basal bodies and most of the axonemal microtubules are resorbed. But the interconnecting ciliary rootlet system of a degenerating epidermal cell withstands resorption to a considerable degree.

"Pulsatile bodies" are not known from any metazoan taxon outside the Acoelomorpha and are hypothesized as an autapomorphy of this taxon. Most probably, the existence of "pulsatile bodies" (the withdrawing of voluminous degenerating epidermal cells into subepidermal digestive tissues) in the Acoelomorpha is correlated with the evolution of a complex, large and stiff interconnected ciliary rootlet system in epidermal cells and the loss of a subepidermal extracellular matrix (basal lamina).

A. Introduction

The existence of so-called "pulsatile bodies" is a well known characteristic of many species of the Acoela and has been reported for the Nemertodermatida as well (WESTBLAD 1950 p. 45; SMITH et al. 1986 p. 19).

Such "pulsatile bodies" are ciliated cells located within the epidermis or within the subepidermal "parenchymal" digestive tissues.

These cells are known for more than a century and were first regarded as parasites (GEDDES 1879 p. 455; von GRAFF 1904–1908 p. 1973; see also DOREY 1965 p. 149). DELAGE (1886) concluded that the "pulsatile bodies" might be flame cells; a possible excretory function of these cells has been discussed by STEINBÖCK (1966 p. 97 and p. 103) and more recently by RIEGER et al. (1991 p. 87). However, most authors who observed "pulsatile bodies" at the light microscopical level regarded these cells to be epidermal replacement cells (e. g. LUTHER 1912 p. 5 and p. 18; WESTBLAD 1940 p. 8, 1942 p. 15, 1948 p. 13; MARCUS 1952 p. 12, p. 103 and p. 127; PAPI 1957 p. 134; REISINGER 1961 p. 8; DÖRJES 1966 p. 190, 1968 for many species, 1971 p. 121; FAUBEL 1974 p. 13; CREZEE 1975 p. 824; EHLERS & DÖRJES 1979 p. 42). On the other hand, BEKLEMISCHEV (1915 p. 159) and MAMKAEV (1965 p. 30, 1967 pp. 33–36, 45–47, 63–65) concluded that these bodies are degenerating epidermal cells being withdrawn from the epidermis and resorbed in subepidermal digestive tissues.

From the light microscope, a clarification of the true nature of these "pulsatile bodies" seems problematically (but see MAMKAEV 1967 for his excellent study).

However, only few studies of "pulsatile bodies" at the EM-level have been done until now. DOREY (1965) concluded, that "pulsatile bodies" are replacement cells, but this author also discussed the possibility, that these cells might be

---

Fig. 1. A, B. Low magnification micrographs from a series of oblique transverse sections of the ▶ cellular epidermis with apical epitheliosomes, parts of the intraepidermal outer longitudinal musculature peripheral to the circular musculature, and an intraepidermal pulsatile body with slightly modified epitheliosomes. Scale bars in all figures: 1 µm.

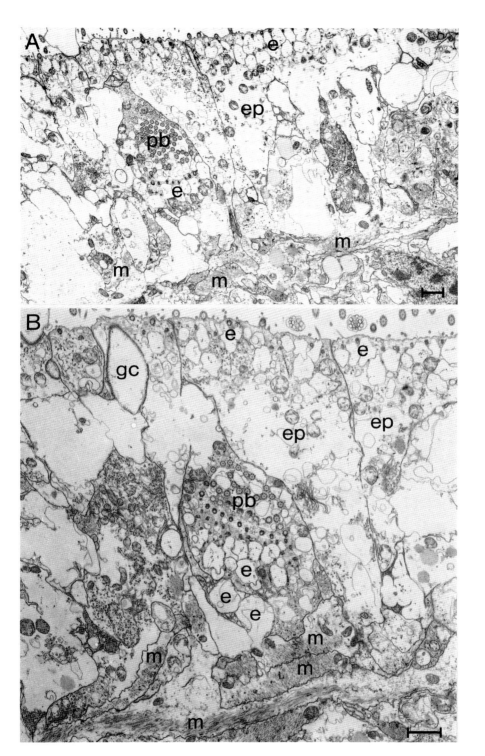

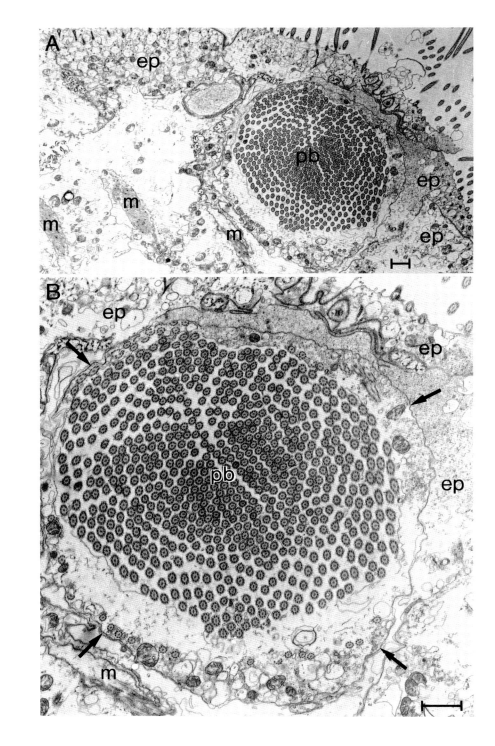

degenerating epidermal cells. TYLER (1984), SMITH et al. (1986) and TYLER et al. (1989) gave strong evidence, from their detailed and excellent studies on the fine structural organization of the Acoelomorpha, that the "pulsatile bodies" are probably damaged epidermal cells being resorbed.

The existence of "pulsatile bodies" in *Anaperus tvaerminnensis* has been noted by LUTHER (1912) in his careful description of this species. An ultrastructural investigation of these cells was undertaken to see whether such cells in wild specimens of *Anaperus t.* are degenerating epidermal cells.

B. Materials and Methods

Anaperus tvaerminnensis (Luther, 1912) was extracted from sediments collected in the eulittoral in front of the old littoral station of the BAH at List/Sylt (North Sea).

Several specimens were prepared for electron microscopy; for EM-processing see EHLERS (1992 a).

C. Results

"Pulsatile bodies" most similar to "normal" epidermal cells were found within the epidermis (Fig. 1). These "pulsatile bodies" are individual cells which display most organelles of neighbouring epithelial epidermal cells. Most of the ciliary axonemes of such a cell situated beneath the epidermal surface are enclosed in one large vacuole surrounded by the cytoplasm of this cell. Several axonemes, some of which do not display a complete pattern of microtubules, exist within this surrounding cytoplasm. The basal bodies of the intact cilia are arranged in distinct longitudinal rows as in apically situated unmodified epidermal cells. Epitheliosomes are also arranged between the ciliary rootlet systems of the "pulsatile bodies".

Most probably, all such slightly modified epidermal cells take an intercellular position, i. e., they do not lie within a vacuole of an epidermal cell but between epidermis cells. However, cell junctions which join the apical regions of epithelial epidermal cells were not observed in any of these slightly modified cells.

◄ Fig. 2. Low magnification (A) and more higher magnification (B) micrograph of an intraepidermal located pulsatile body with cross-sectioned cilia. Arrows point to the outer mostly complete (but lacking cell junctions) cell membrane of the pulsatile body. Cytoplasm of the body with intact but also partly resorbed ciliary axonemes.

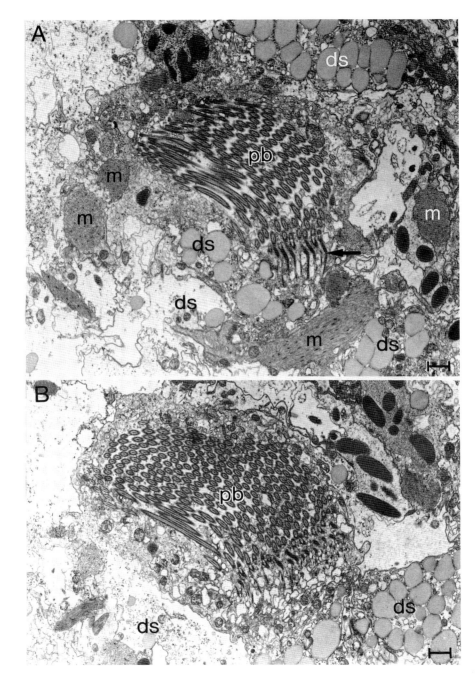

Sometimes, the cell membranes of these cells are partly interrupted and the same is true for the adjacent functioning epidermal cell (Fig. 1 B), so that the cytoplasms of both cells make contact.

Whereas the cilia of the "pulsatile bodies" figured in Fig. 1 project peripherally like "normal" epidermal cells, those of the "pulsatile body" figured in Fig. 2 run parallel to the epidermal surface, i. e., here the whole cell with its cilia has changed its position for 90° in comparison with the neighbouring epidermal cells. The existence of ciliary axonemes with incomplete numbers of microtubules can be clearly seen in the peripheral cytoplasm of such tilted cells (Fig. 2 B).

In *Anaperus t.*, "pulsatile bodies" can be found in the subepidermal digestive systems as well. Apparently, the ciliated cells lie between cells when located in the more peripheral cellular "parenchymal" digestive tissue. But the cells become enclosed in vacuoles when reaching the central syncytial areas of the digestive system (Figs. 3 A, 4–7). Here, the "pulsatile bodies" undergo considerable cytological changes.

The numbers of the intact ciliary axonemes are lowered in cells located in the peripheral digestive system (Figs. 3, 4). Most probably, a considerable number of those ciliary axonemes originally located at the periphery of an intact epidermal cell or of an intraepidermal "pulsatile body", respectively, have been resorbed in the cytoplasm of the bodies. The remaining intact ciliary axonemes still display a distal shelf (Fig. 4) and the rootlet systems of these more centrally located cilia are complete and interconnected (Figs. 3 A, 4). At this stage of autolysis, the membranes of the vacuole enclosing the intact ciliary axonemes become partly fragmented and the mitochondria within the cytoplasm of the bodies are in different stages of autolysis.

All the ciliary axonemes of a "pulsatile body" are no longer intact when the surrounding membranes of the vacuole are destroyed (Fig. 5). In such a stage of degeneration, only remnants of microtubules can be seen in the axonemes. Furthermore, all the basal bodies of the cilia have been resorbed, so that the still complete rootlet systems do not make contact with the axonemal remnants (Fig. 5).

The most advanced stages of degeneration which have been found in *Anaperus t.* are figured in Figs. 6 and 7. Here, most of the cytoplasm of the bodies has been resorbed. And nearly all of the ciliary axonemes, which lack any microtubules, have been fragmented, but parts of the rootlet systems still exist.

◄ Fig. 3. A, B. Pulsatile body at the boundery between the peripheral parts (right side) and the central parts (left side) of the digestive system; cytoplasm of the body partly (left side in A) already enclosed in this system. Ciliary axonemes (enclosed in a vacuole) and ciliary rootlets (arrow in A) still intact.

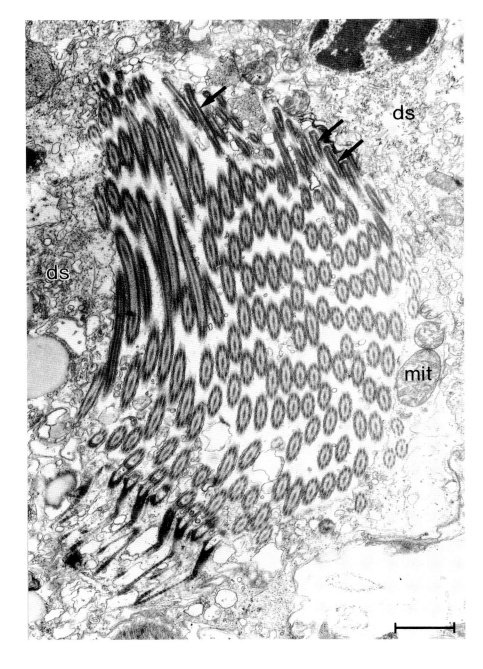

"Pulsatile bodies" in *Anaperus* 303

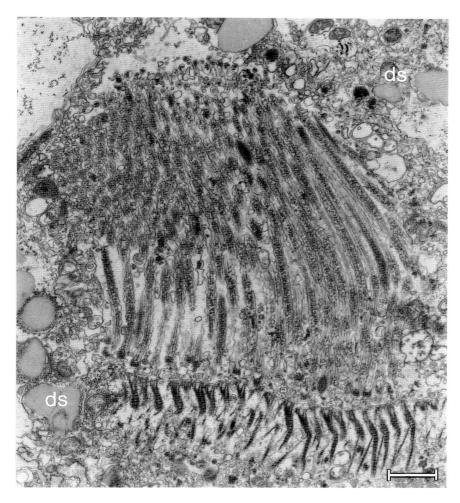

Fig. 5. Pulsatile body (within the digestive system) with degenerated ciliary axonemes, but still intact interconnected ciliary rootlet systems.

◄ Fig. 4. Pulsatile body (micrograph comparable to fig. 3 A) partly enclosed in the digestive system. Arrows point to the still intact distal shelfs of many ciliary axonemes lying inside a vacuole with a partly fragmented membrane (upper half of the figure); ciliary rootlet systems still complete and interconnected.

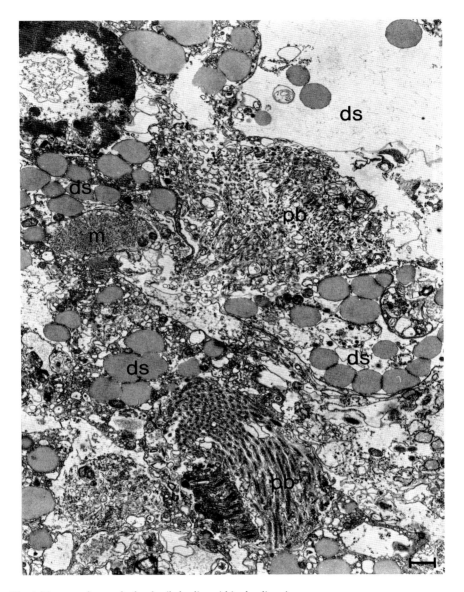

Fig. 6. Two mostly resorbed pulsatile bodies within the digestive system.

Fig. 7. In A remnants of a pulsatile body being resorbed but with ciliary rootlet systems still visible. In B another pulsatile body in the process of autolysis.

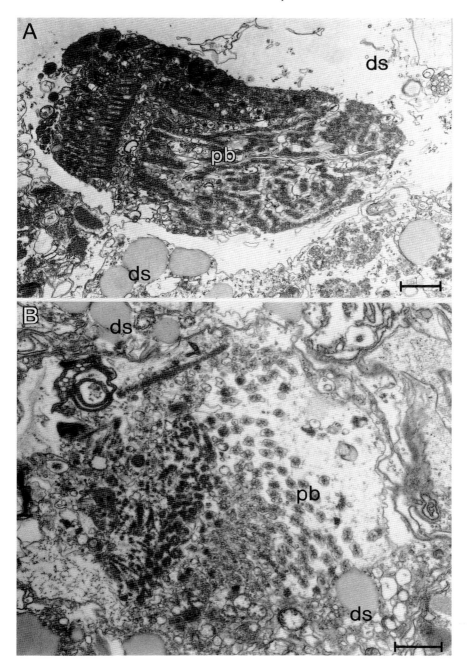

D. Discussion

There is no doubt, that the "pulsatile bodies" of *Anaperus tvaerminnensis* are degenerating epidermal cells, which are withdrawn from the epidermis and which become enclosed in vacuoles of the central syncytial digestive system for resorption.

Most probably, the cilia of an epidermal cell being withdrawn from the surface become enclosed in a vacuole being formed by the cell itself. But there is the possibility as well, that at least parts of the cytoplasm covering the ciliary axonemes are derived from neighbouring epidermal cells. Such a hypothesis is based on observations where a breakdown of cell membranes of both the "pulsatile body" and an adjacent epidermal cell were found. In any case, cells being withdrawn from the surface do not display zonulae adhaerentes and septate junctions. However, there is the possibility that gap junctions between intraepidermal bodies and neighbouring epidermal cells may exist. But the mechanisms responsible for the migration of the "pulsatile bodies" from the epidermis to the central digestive syncytium can not be explained on the basis of the available informations.

From the observed stages of degeneration it can be concluded that resorption of the most peripheral cilia of a degenerating epidermis cell occur within the cytoplasm as soon as a cell becomes withdrawn whereas the more centrally located cilia are resorbed in subepidermal tissues.

That means, the centrally located cilia are capable for pulsation within the vacuoles as long as such a degenerating cell is located within the epidermis or the more peripheral subepidermal body regions. However, cells which become enclosed in the central digestive system and which show advanced stages of autolysis, do not possess cilia capable for pulsation. In the light microscope, such more degenerated cells are hardly to detect.

This circumstance explains why several authors have failed to detect "pulsatile bodies" in subepidermal tissues with the light microscope. And the controversial discussions about the existence or the lack of nuclei in such cells can also be explained by the different degrees of degeneration.

And, of course, it is not surprising now why several of those authors, who regarded the "pulsatile bodies" as epidermal replacement cells, have wondered why they failed to find any precursors of these cells or any cells in the process of early differentiation.

The new observations on "pulsatile bodies" in *Anaperus tvaerminnensis* confirm the earlier conclusions presented by BEKLEMISCHEV (1915) and especially by MAMKAEV (1967). MAMKAEV (l. c., fig. 6) gave a detailed light microscopical analysis of degeneration for "pulsatile bodies" in the acoel *Haploposthia opisthor-*

chis which correspond well with the degeneration processes seen in *Anaperus tvaerminnensis*.

As stated already by TYLER (1984), SMITH et al. (1986) and TYLER et al. (1989), "pulsatile bodies" result from damaged cells, but such bodies can be found in the uninjured epidermis of wild specimens as well. Here, the "pulsatile bodies" most probably represent epidermal cells being replaced by new differentiated cells. But many specimens of different species of the Acoela (and Nemertodermatida as well) investigated at the EM-level in our laboratory lack "pulsatile bodies". Apparently, replacement of epidermal cells in uninjured specimens is not a permanent process. In *Anaperus t.*, no indication of an excretory function of "pulsatile bodies" has been found.

The existence of "pulsatile bodies" has been reported for the Acoela and by WESTBLAD (1950) for *Meara stichopi*, a species of the Nemertodermatida, as well. STEINBÖCK (1967) questioned the observations of WESTBLAD (l. c.), but SMITH et al. (1986) observed "pulsatile bodies" in a species of the taxon *Nemertoderma*. Hence, this characteristic can be hypothesized as an autapomorphy of the taxon Acoelomorpha (EHLERS 1992 b).

"Pulsatile bodies" are not known from any metazoan taxon outside the Acoelomorpha. In particular, no observations on ciliated cells being withdrawn from the surface and migrating into subepidermal tissues are available for other plathelminth taxa, the Catenulida and the Rhabditophora, respectively. Most probably, the existence of "pulsatile bodies" in the Acoelomorpha is correlated with the existence of a discontinuous subepidermal matrix in the Nemertodermatida and the total lack of any matrices in the Acoela. With the – secondary – absence of such matrices, especially basiepidermal layers, the migration of voluminous differentiated (ciliated) epidermal cells towards the digestive system became possible. In Plathelminthes with subepidermal matrices, cells like differentiating epidermis cells diminuish their diameters considerably when penetrating such a matrix (cf. EHLERS 1985, table 15). In contrast to the Catenulida and the Rhabditophora, the differentiated (ciliated) epidermal cells of the Acoelomorpha, especially those of the Acoela, do not change their sizes because of the existence of the stiff interconnected ciliary rootlet systems, which withstand degeneration processes to a considerable degree. That means, the unique phenomenon of the "pulsatile bodies" seems to be correlated with the evolution of an interconnected epidermal ciliary rootlet system and the loss of a subepidermal basal matrix in the Acoelomorpha.

Acknowledgements

I would like to thank Mrs. E. Hildenhagen-Brüggemann for making the EM-preparations and for taking the electron micrographs and Mr. B. Baumgart for his help with the figures. Financial support was provided by the Akademie der Wissenschaften und der Literatur, Mainz.

Zusammenfassung

Aus dem Freiland entnommene Individuen von *Anaperus tvaerminnensis* (Luther, 1912) besitzen sogenannte „pulsierende Körper". Diese treten intraepidermal wie auch subepidermal auf. Es handelt sich hierbei nicht um „Regenerationskörper", sondern um degenerierende Epidermiszellen, die aus der epithelialen Epidermis zurückgezogen werden, im zentralen Verdauungssyncytium in Vakuolen eingeschlossen werden und hier der Autolyse unterliegen.

Die ciliären Axonemata der zurückgezogenen Epidermiszellen werden während der intraepidermalen Passage nur partiell resorbiert, zum größeren Teil aber intakt in eine Vakuole eingeschlossen. Die Möglichkeit zum Cilienschlag ist beendet, wenn die Zelle im verdauenden Körperbereich angelangt ist: hier bricht die Vakuolenmembran um die Axonemata auf und alle Basalkörper und letztlich auch die meisten Mikrotubuli aller Axonemata werden resorbiert. Von allen Strukturen einer solchen Zelle bleibt das komplizierte Wurzelsystem, das alle Cilien einer Epidermiszelle miteinander verbindet, am längsten erhalten.

Die Existenz solcher „pulsierender Körper" ist auf das Taxon Acoelomorpha beschränkt und bildet eine Autapomorphie dieses Taxons. Sehr wahrscheinlich hat sich dieser einmalige Vorgang des Zurückziehens ganzer voluminöser degenerierender Epidermiszellen im Zusammenhang mit der Evolution des starren Verbundsystems der komplexen epidermalen Cilienwurzeln und dem Verlust einer, die Wanderung sonst hemmenden, subepidermalen Basallamina herausgebildet.

Abbreviations

ds	digestive system		m	musculature
e	epitheliosome(s)		mit	mitochondrion
ep	epidermal cell(s)		pb	pulsatile body
gc	neck of epidermal gland cell			

References

BEKLEMISCHEV, W. N. (1915): Sur les turbellariés parasites de la côte Mourmanne. I. Acoela. Trav. Soc. Nat. Pétrograd **43**, 103–172 (In Russian, French resume).

CREZEE, M. (1975): Monograph of the Solenofilomorphidae (Turbellaria: Acoela). Int. Revue ges. Hydrobiol. **60**, 769–845.

DELAGE, Y. (1886): Etudes histologiques sur les Planaires Rhabdocoeles Acoeles (*Convoluta schultzii* O. Sch.). Arch. Zool. exp. gén. **4**, 109–144.

DÖRJES, J. (1966): *Paratomella unichaeta* nov. gen. nov. spec., Vertreter einer neuen Familie der Turbellaria Acoela mit asexueller Fortpflanzung durch Paratomie. Veröffentl. Inst. Meeresforsch. Bremerhaven Sonderband **2**, 187–200.

– (1968): Die Acoela (Turbellaria) der deutschen Nordseeküste und ein neues System der Ordnung. Z. zool. Syst. Evolut.-forsch. **6**, 56–452.

– (1971): Monographie der Proporidae und Solenofilomorphidae (Turbellaria Acoela). Senckenbergiana biol. **52**, 113–137.

DOREY, A. E. (1965): The organization and replacement of the epidermis in acoelous turbellarians. Quart. J. micr. Sci. **106**, 147–172.

EHLERS, U. (1985): Das phylogenetische System der Plathelminthes. G. Fischer, Stuttgart, New York, 317 p.

– (1992 a): Dermonephridia – modified epidermal cells with a probable excretory function in *Paratomella rubra* (Acoela, Plathelminthes). Microfauna Marina **7**, 253–264.

– (1992 b): On the fine structure of *Paratomella rubra* Rieger & Ott (Acoela) and the position of the taxon *Paratomella* Dörjes in a phylogenetic system of the Acoelomorpha (Plathelminthes). Microfauna Marina **7**, 265–293.

EHLERS, U. & J. DÖRJES (1979): Interstitielle Fauna von Galapagos. XXIII. Acoela (Turbellaria). Mikrofauna Meeresboden **72**, 1–75.

FAUBEL, A. (1974): Die Acoela (Turbellaria) eines Sandstrandes der Nordseeinsel Sylt. Mikrofauna Meeresboden **32**, 1–58.

GEDDES, P. (1879): Observations on the physiology and histology of *Convoluta schultzii*. Proceed. Roy. Soc. London **28**, 449–457.

GRAFF, L. v. (1904–1908): Acoela und Rhabdocoelida. Plathelminthes III. Turbellaria. In: Bronn's Klassen und Ordnungen des Thierreichs Vol. IV, Abt. Ic. Winter'sche Verlagshandlung, Leipzig, 1733–2599.

LUTHER, A. (1912): Studien über acöle Turbellarien aus dem finnischen Meerbusen. Acta Soc. Fauna Flora Fennica **36**, No. 5, 1–60.

MAMKAEV, Yu. V. (1965): Etude morphologique d'*Actinoposthia beklemischevi* n. sp. (Turbellaria Acoela). Cah. Biol. Mar. **6**, 23–50.

– (1967): Essays on the morphology in acoelous Turbellaria. Trudy Zool. Inst. Akad. Nauk SSSR **44**, 26–108 (In Russian).

MARCUS, E. (1952): Turbellaria Brasileiros (10). Bol. Fac. Fil. Cienc. Letr. Univ. Sao Paulo, Zool. **17**, 5–187.

PAPI, F. (1957): Sopra un nuovo Turbellario arcooforo di particolare significato filetico e sulla posizione della fam. Hofsteniidae nel sistema dei Turbellari. Publ. Staz. Zool. Napoli **30**, 132–148.

REISINGER, E. (1961): Allgemeine Morphologie der Metazoen. Morphologie der Coelenteraten, acoelomaten und pseudocoelomaten Würmer. Fortschr. Zool. **13**, 1–82.

RIEGER, R. M., S. TYLER, J. P. S. SMITH III & G. E. RIEGER (1991): Platyhelminthes: Turbellaria. In: Microscopic anatomy of invertebrates (Treatise ed.: F. W. HARRISON), Vol. 3, Platyhelminthes and Nemertinea. Eds.: F. W. HARRISON & B. J. BOGITSH. Wiley-Liss, New York, 7–140.

SMITH III, J. P. S., S. TYLER & R. RIEGER (1986): Is the Turbellaria polyphyletic? Hydrobiologia **132**, 13–21.

STEINBÖCK, O. (1966): Die Hofsteniiden (Turbellaria acoela). Grundsätzliches zur Evolution der Turbellarien. Z. zool. Syst. Evolut.-forsch. **4**, 58–195.

— (1967): Regenerationsversuche mit *Hofstenia giselae* Steinb. (Turbellaria Acoela). Roux' Arch. Entwicklungsmech. **158**, 394–458.

TYLER, S. (1984): Turbellarian platyhelminths. In: Biology of the integument, Vol. 1, Invertebrates. Eds.: J. BEREITER-HAHN, A. G.MATOLTSY & K. S. RICHARDS. Springer, Berlin, 112–131.

TYLER, S., J. K. GRIMM & J. P. S. SMITH III (1989): Dynamics of epidermal wound repair in acoel turbellarians – the role of pulsatile bodies. Am. Zool. **29**, 115 A.

WESTBLAD, E. (1940): Studien über skandinavische Turbellaria Acoela. I. Ark. Zool. **32 A**, No. 20, 1–28.

— (1942): Studien über skandinavische Turbellaria Acoela. II. Ark. Zool. **33 A**, No. 14, 1–48.

— (1948): Studien über skandinavische Turbellaria Acoela. V. Ark. Zool. **41 A**, No. 7, 1–82.

— (1950): On *Meara stichopi* (Bock) Westblad, a new representative of Turbellaria Archoophora. Ark. Zool. Ser. 2, **1**, 43–57.

Prof. Dr. Ulrich Ehlers
II. Zoologisches Institut und Museum der Universität Göttingen
Berliner Straße 28, D-3400 Göttingen

No mitosis of differentiated epidermal cells in the Plathelminthes: mitosis of intraepidermal stem cells in *Rhynchoscolex simplex* Leidy, 1851 (Catenulida)

Ulrich Ehlers

Contents

Abstract	311
A. Introduction	312
B. Materials and Methods	313
C. Results	313
D. Discussion	315
Acknowledgements	320
Zusammenfassung	320
Abbreviations in the figures	320
References	320

Abstract

The monolayered uninjured epidermis of *Rhynchoscolex simplex* consists of ciliated epidermal cells and several accumulations of undifferentiated cells (stem cells) wedged between the basal portions of the differentiated epidermal cells. These stem cells undergo intraepidermal mitoses; mitoses of differentiated cells do not occur.

A stem cell in a M-phase is much more voluminous than a stem cell in a G-phase. Most probably, a mitosis of such a large cell leads to the genesis of two precursors of new epidermal cells. Ciliogenesis starts during metaphase with the genesis of a cluster of procentrioles in both clouds with centrosomal materials at the spindle poles.

A. Introduction

There is strong evidence that differentiated epidermal cells in Plathelminthes do not divide. New epidermal cells differentiate from stem cells which are capable of mitosis (for discussions see e. g. BAGUNA et al. 1989; EHLERS 1985; PALMBERG 1990; RIEGER et al. 1991).

Undifferentiated cells, which appear to be precursors of new epidermal cells, have been found in the peripheral body regions beneath the body wall musculature in many taxa of the Rhabditophora and the Catenulida as well and have been observed to immigrate into the epidermis. At the EM-level, such migrating cells show a cluster of centrioles which are thought to represent precursors of basal bodies for ciliogenesis. Immigrated stem cells (or progenitors of epidermal cells) have been found in basiepidermal regions of different plathelminth taxa as well.

The taxon Catenulida is the only plathelminth taxon for which intraepidermal mitoses in adult limnic specimens have been described (e. g. REISINGER 1924; STERN 1925; MARCUS 1945; PULLEN 1957). STERN (l. c.) concluded, that the mi-

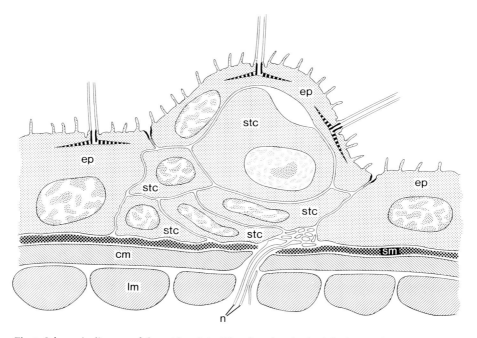

Fig. 1. Schematic diagram of the epidermis in *Rhynchoscolex simplex*. The intraepidermal stem cells, one of which is in the beginning of mitosis, are covered by the monolayered differentiated epidermal cells.

totic figures occurred within differentiated epidermal cells. On the other hand, PULLEN (l. c.) described cell divisions from intraepidermal stem cells ("interstitial cells") and denied the occurrence of mitoses in differentiated epidermis cells. All these older observations are based on light microscopical investigations, an EM-analysis of intraepidermal mitoses in the limnic Catenulida is lacking.

REISINGER (1924, p. 22) described intraepidermal nuclei surrounded by a thin layer of plasma in the limnic catenulid *Rhynchoscolex simplex,* this author also observed mitoses within the epidermis of *R. simplex.*

An EM-analysis of the epidermis of this species should clarify whether differentiated epidermal cells or intraepidermal stem cells undergo mitoses in the limnic Catenulida.

B. Materials and Methods

Rhynchoscolex simplex Leidy, 1851 was extracted from sediments collected at the small river Lachte near Celle (Lower Saxonia) by Dr. R. Altmüller. Several specimens were prepared for electron microscopy; for EM-processing see EHLERS (1992).

C. Results

The epidermis of *Rhynchoscolex simplex* is a monolayered cellular and ciliated epithelium. All the basal bodies of the locomotory cilia are situated in pits of the cell surface and provided with a rostral and a caudal rootlet. In general, the ciliation is of a low density. A prominent surface coat (glycocalyx) covers the epidermis and extends between the microvilli. All the nuclei of the epidermal cells are intraepithelial, "insunk" pericarya were not observed.

The epidermal cells rest on an inconspicuous lamina-densa-like extracellular matrix, which separates the epidermis from the outer circular and inner longitudinal musculature. Processes of nerve cells penetrate the matrix layer, i. e., the peripheral nervous system of *Rhynchoscolex simplex* consists of subepidermal and of basiepidermal parts as well.

Besides the ciliated epidermal cells, undifferentiated cells can be found in several regions of the epidermis. These cells, which are much smaller than the differentiated cells and which display a high nucleus/cytoplasm ratio, always take a basiepidermal position. Chromatoid bodies are present but difficult to see because of the abundant ribosomes of these cells which are always covered by differentiated epidermal cells (Figs. 1–3).

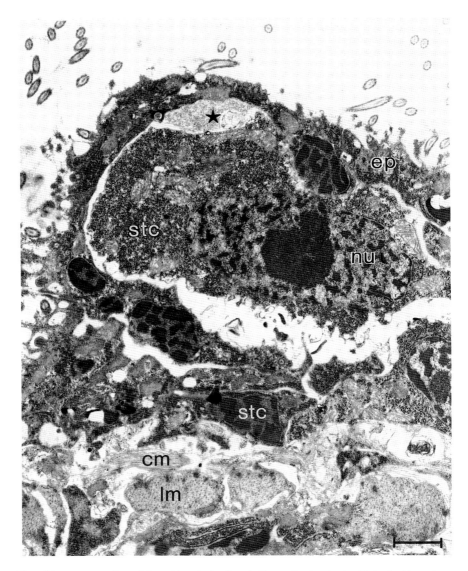

Fig. 2. Transverse section of the epidermis (in view similar to that in Fig. 1) with basiepidermal stem cell(s) and with a more peripherally located enlarged stem cell with an enlarged nucleus showing a marked nucleolus and scattered heterochromatin still surrounded by the nucleus membrane. Differentiated epidermis cells partly separated from the enlarged stem cell by intercellular spaces with flocculent materials (asterisk). Scale bars in Figs. 2–5: 1 µm.

The undifferentiated cells, which represent stem cells, undergo mitoses within the epidermis. At least in an early stage of the prophase (but perhaps already in the S- or G_2-phases) the cells become much more voluminous (Figs. 1–2), so that the epidermis becomes protuberant. At this stage, the enlarged stem cell and the overlying differentiated epidermal cells can be partly separated from each other by irregular intercellular spaces filled with flocculent materials (Fig. 2). Stem cells in such an early stage of the M-phase display an enlarged nucleus with a prominent nucleolus and with more scattered heterochromatin still surrounded by the nucleus membrane.

Cells in a more advanced stage of mitosis (metaphase) are of the same enlarged sizes and take the same positions within the epidermis as in the beginning of the M-phase (Fig. 3 A). Surprisingly, one stem cell in a stage typical of the metaphase (with chromosomes in the metaphase plate) shows one chromatid already located near one spindle pole (Fig. 3 B).

During the metaphase stage, numerous astral and polar microtubules exist in the vicinity of both clouds with centrosomal materials in the two spindle poles. Not only the two centrioles of a diplosom, but also a cluster of procentrioles have been found in most clouds of centrosomal material (Figs. 4–5). The diplosomal centrioles lack a cartwheel-like structure which is typical of basal bodies of epidermal cilia in the Catenulida (cf. EHLERS 1986, fig. 5 B).

In principle, only one stem cell of a local intraepidermal stem cell accumulation can be seen in the M-phase whereas the other stem cells are in a G-phase (Figs. 2–4). Apparently, all the intraepidermal stem cells lack zonulae adhaerentes, septate junctions and tight junctions whereas the differentiated epidermis cells are joined apically by zonulae adhaerentes and septate junctions (Figs. 1–2).

D. Discussion

The fine structural organization of the epidermis with its differentiated ciliated cells corresponds with the epidermis of other limnic catenulids like *Catenula* (SOLTYNSKA et al. 1976; MORACZEWSKI 1981; EHLERS 1985) and *Stenostomum* (PEDERSEN 1983). That means, the earlier reports of MARCUS (1945) on a syncytial organization of the epidermis in *Rhynchoscolex simplex* and of REISINGER (1924) on the existence of epitheliomuscular cells must be corrected.

As in other plathelminths, mitoses in differentiated epidermal cells of *R. simplex* could not be observed. The only cells capable of mitosis are stem cells. In *R. simplex*, several intraepidermal accumulations of stem cells have been found in different body regions. These cells take a basiepidermal position and are always overlayered by the differentiated ciliated epidermal cells as described for

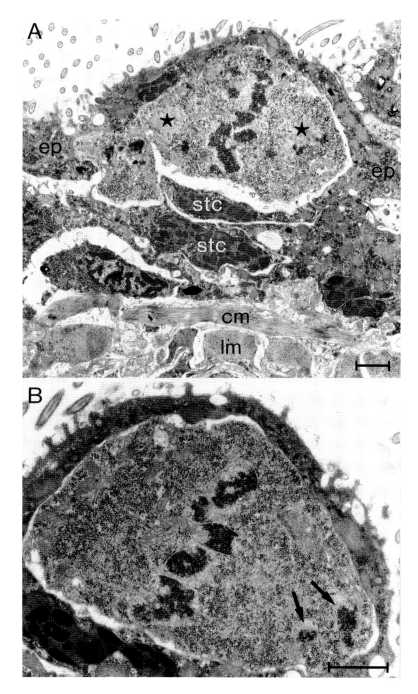

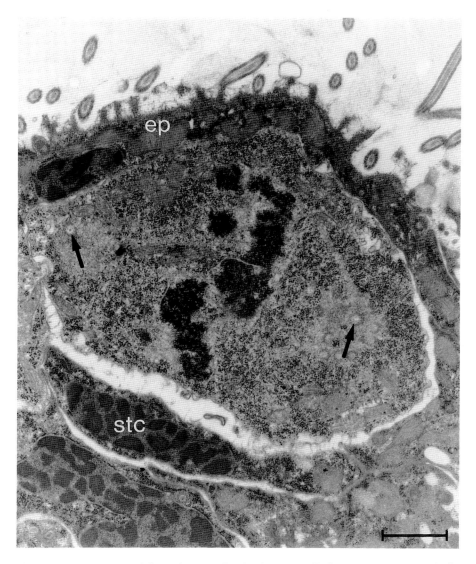

Fig. 4. Transverse section of the epidermis with a dividing stem cell (the same as in Fig. 3 A) in the stage of metaphase. In both regions of the spindle poles, a cluster of procentrioles exists besides a diplosom (arrows point to one centriole of each diplosom).

◄ Fig. 3. Transverse sections of the epidermis. In A an enlarged stem cell in the stage of metaphase (the two spindle poles are marked by asterisks) and two other stem cells (with blocky heterochromatin) in the G_1-phase. In B an enlarged stem cell in the stage of (late?) metaphase but with one chromatid (arrows) located near the right spindle pole already.

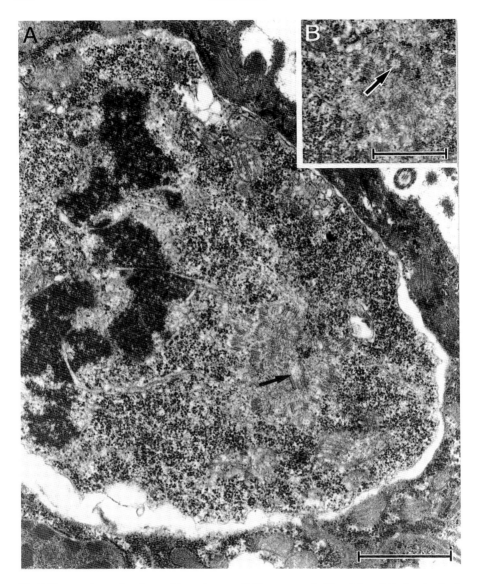

Fig. 5. Two sections of an intraepidermal stem cell (the same as in Figs. 3 A and 4) showing both centrioles of one spindle pole region: one centriole (arrow in A) in longitudinal, the second centriol (arrow in B) in transverse section. Both centriols in diplosomal configuration are surrounded by many procentrioles concentrated in the cloud of centrosomal material from which astral and polar microtubules spread out.

other limnic Catenulida from the light microscope (e. g. PULLEN 1957; BORKOTT 1970).

Stem cells in subepidermal as well as in basiepidermal positions have been reported for the limnic taxa *Stenostomum* and *Catenula* (e. g. MARCUS 1945; BORKOTT 1970; SOLTYNSKA et al. 1976; MORACZEWSKI 1981; PEDERSEN 1983). From the electron microscope, such intraepidermal or subepidermal cells have been described as cells containing many centrioles (SOLTYNSKA et al. 1976; MORACZEWSKI 1981; PEDERSEN 1983) and have been interpreted as precursors of epidermal cells. Mitoses of stem cells in the Catenulida have not been documented from the EM-level until now.

From the new findings it becomes obvious that stem cells lacking clusters of centrioles (or procentrioles) also exist. Such cells undergo mitosis in *Rhynchoscolex simplex*. At least in the beginning of the M-phase, these cells are much more voluminous than stem cells in a G-phase. Most probably, these larger cells correspond with intraepidermal cells already observed by REISINGER (1924, p. 23) who interpreted these large cells as degenerative germ cells. REISINGER (l. c.) believed that a special kind of intraepidermal nuclei surrounded by a thin layer of cytoplasm (apparently stem cells in the G-phase as described here) are the precursors of female germ cells which migrate into subepidermal regions. From the new EM-findings it seems unlikely, that intraepidermal stem cells differentiate to female germ cells. Apparently, the intraepidermal stem cells of *Rhynchoscolex simplex* are precursors of epidermal cells.

EM-observations on ciliogenesis in other animal taxa have shown, that a kinetosomal mode of genesis for putative basal bodies is a common feature but never started during the M-phase of stem cells nor of blastomeres, respectively (cf. lit. in TYLER 1981, 1984). It seems likely, that a mitosis of a large intraepidermal stem cell in *Rhynchoscolex simplex* leads to the genesis of two precursors of epidermal cells each equipped with a cluster of centrioles.

As stated already, the taxon Catenulida is the only plathelminth taxon for which intraepidermal mitoses are known. But mitoses of intraepidermal stem cells are known for other taxa outside the Plathelminthes, for example in the Cnidaria. Apparently, the existence of intraepidermal stem cells which undergo mitoses is a plesiomorphic characteristic for the Catenulida. This characteristic no longer exists in all other plathelminths, the Euplathelminthes.

However, the early genesis of centrioles (the beginning of ciliogenesis) in the M-phase of intraepidermal stem cells may represent a unique characteristic of *Rhynchoscolex simplex*.

Acknowledgements

I thank Mrs. E. Hildenhagen-Brüggemann for making the EM-preparations and for taking the electron micrographs and Mr. B. Baumgart for the realization of the drawing. Financial support was provided by the Akademie der Wissenschaften und der Literatur, Mainz.

Zusammenfassung

Die einschichtige unverletzte Epidermis von *Rhynchoscolex simplex* besteht aus cilientragenden Epidermiszellen und einigen Anhäufungen von undifferenzierten Zellen (Stammzellen), eingekeilt zwischen den basalen Bereichen der differenzierten Epidermiszellen. Diese Stammzellen teilen sich mitotisch; bei differenzierten Zellen tritt eine Mitose nicht auf.

In der M-Phase ist eine Stammzelle wesentlich voluminöser als während der G-Phase. Sehr wahrscheinlich führt die Teilung einer solch großen Zelle zur Bildung von zwei prospektiven Epidermiszellen. Die Ciliogenese beginnt bereits während der Metaphase mit der Genese von Procentriolen an beiden Spindelpolen.

Abbreviations

cm	circular musculature	nu	nucleus of a stem cell
ep	epidermis cell(s)	sm	subepidermal extracellular matrix
lm	longitudinal musculature	stc	stem cell(s)
n	peripheral nervous system		

References

BAGUNA, J., E. SALO & C. AULADELL (1989): Regeneration and pattern formation in planarians. III. Evidence that neoblasts are totipotent stem cells and the source of blastema cells. Development 107, 77–86.

BORKOTT, H. (1970): Geschlechtliche Organisation, Fortpflanzungsverhalten und Ursachen der sexuellen Vermehrung von *Stenostomum sthenum* nov. spec. (Turbellaria, Catenulida). Mit Beschreibung von 3 neuen *Stenostomum*-Arten. Z. Morph. Tiere 67, 183–262.

EHLERS, U. (1985): Das phylogenetische System der Plathelminthes. G. Fischer, Stuttgart, New York, 317 p.

– (1986): Comments on a phylogenetic system of the Platyhelminthes. Hydrobiologia 132, 1–12.

– (1992): Dermonephridia – modified epidermal cells with a probable excretory function in *Paratomella rubra* (Acoela, Plathelminthes). Microfauna Marina 7, 253–264.

MARCUS, E. (1945): Sobre Catenulida brasileiros. Bol. Fac. Fil. Cienc. Letr. Univ. Sao Paulo, Zool. 10, 3–133.
MORACZEWSKI, J. (1981): Fine structure of some Catenulida (Turbellaria, Archoophora). Zool. Pol. 28, 367–415.
PALMBERG, I. (1990): Stem cells in microturbellarians. An autoradiographic and immunocytochemical study. Protoplasma 158, 109–120.
PEDERSEN, K. J. (1983): Fine structural observations on the turbellarians *Stenostomum* sp. and *Microstomum lineare* with special reference to the extracellular matrix and connective tissue systems. Acta Zool. (Stockh.) 64, 177–190.
PULLEN, E. W. (1957): A histological study of *Stenostomum virginianum*. J. Morph. 101, 579–621.
REISINGER, E. (1924): Die Gattung *Rhynchoscolex*. Z. Morph. Ökol. Tiere 1, 1–37.
RIEGER, R. M., S. TYLER, J. P. S. SMITH III & G. E. RIEGER (1991): Platyhelminthes: Turbellaria. In: Microscopic anatomy of invertebrates (Treatise ed.: F. W. HARRISON), Vol. 3 Platyhelminthes and Nemertinea. Eds.: F. W. HARRISON & B. J. BOGITSH. Wiley-Liss, New York, 7–140.
SOLTYNSKA, M. S., B. MROCZKA & J. MORACZEWSKI (1976): Ultrastructure of epidermis in Turbellaria from the family Catenulida (Archoophora). J. Submicr. Cytol. 8, 293–301.
STERN, C. (1925): Die Mitose der Epidermiskerne von *Stenostomum*. Z. Zellforsch. 2, 121–128.
TYLER, S. (1981): Development of cilia in embryos of the turbellarian *Macrostomum*. Hydrobiologia 84, 231–239.
– (1984): Ciliogenesis in embryos of the acoel turbellarian *Archaphanostoma*. Trans. Am. Microsc. Soc. 103, 1–15.

Prof. Dr. Ulrich Ehlers
II. Zoologisches Institut und Museum der Universität Göttingen
Berliner Straße 28, D-3400 Göttingen

The larval protonephridium of *Stylochus mediterraneus* Galleni (Polycladida, Plathelminthes): an ultrastructural analysis

Cornelia Wenzel, Ulrich Ehlers and **Alberto Lanfranchi**

Contents

Abstract	323
A. Introduction	324
B. Materials and Methods	324
C. Results	325
D. Discussion	335
Acknowledgements	338
Zusammenfassung	339
Abbreviations	339
References	340

Abstract

The larva of *Stylochus mediterraneus* Galleni, 1976 has one pair of unramified protonephridia, each composed of two cells: one terminal cell and one tubule cell. Both protonephridia extend throughout the body length beneath the subepidermal basal lamina and open caudally through intraepidermal nephropori which are formed by the distal areas of the tubule cells.

Each terminal cell is splitted in its proximal region, the longitudinal cleft extends on one side to the adjacent tubule cell, but is partly closed on the other side of the terminal cell. Up to 30 cilia and many microvilli of varying diameters arise from the central luminal cytoplasm of this cell and project into the outleading lumen. The filter areas of the protonephridium are formed by interdigitating microvilli-like cytoplasmic differentiations which are located in the regions of the cleft and which display an extracellular ECM-diaphragm.

The ciliated tubule cell, which lacks a longitudinal cell gap, displays conspicuous folded inner membranes when penetrating the basal lamina.

An additional cell has been found in a more developed protonephridium. This cell made contact with the tubule cell near the nephroporus and might represent a prospective new nephroporus or tubule cell. Most probably, the larval nephridium with its two cells developes to a multi-cellular and ramified protonephridium in adult polyclads by adding more cells to the already existing nephridial organ.

A. Introduction

The fine structural organization of the protonephridia in many species of the plathelminth taxa Catenulida, Macrostomida and Neoophora is well-known, especially the terminal regions with the filter or the weir (lit. in EHLERS 1985, 1989) whereas any detailed ultrastructural study of the protonephridia in the Polycladida is lacking.

LANG (1884) figured and described parts of a protonephridial system of the polyclad *Thysanozoon brocchii*. Short notices are given by i. a. BRESSLAU (1928–33), HYMAN (1951) and PRUDHOE (1985), but, as stated by PRUDHOE (l. c.) "this system in polyclads is little known".

This is especially true for the systems in the larvae of polyclads. RUPPERT (1978) figured the extensions of the larval nephridia in a Götte's larva, but presented only one EM-figure. Another EM-Figure is given by LACALLI (1983) for a Müller's larva.

The protonephridium in a Götte's larva of *Stylochus mediterraneus* was investigated to see whether a larval protonephridium in a polyclad is a completely differentiated organ or not.

B. Materials and Methods

Larvae of *Stylochus (Imogine) mediterraneus* Galleni, 1976 were obtained from adult specimens collected in the Ligurian Sea and fixed at the three-eyed stage (see LANFRANCHI et al. (1981) for further details). For fixations and embedments see also LANFRANCHI et al. (1981).

Serial ultrathin sections (transverse and longitudinal sections) were stained with uranyl acetate and lead citrate and analyzed with Zeiss EM9 and EM10B electron microscopes.

C. Results

In the three-eyed stage, the Götte's larva of *Stylochus mediterraneus* possesses one pair of protonephridia which extend laterally along the longitudinal axis of the specimens (Fig. 1). Both protonephridia originate ventrolaterally at the level of the unpaired ventral lobe and terminate in the posterior body region at the level of the two larger ventral lobes (Fig. 1).

Each protonephridium is about 95 μm in length (length of larva: 160 μm) and runs straight just beneath the subepidermal basal lamina (Figs. 3, 5 C, 8 A). Only two cells occur: an anterior terminal cell (about 30 μm in length) and a posterior canal or tubule cell (about 65 μm in length), which penetrates the basal lamina and forms an intraepidermal nephroporus (Fig. 2). The transition region of both

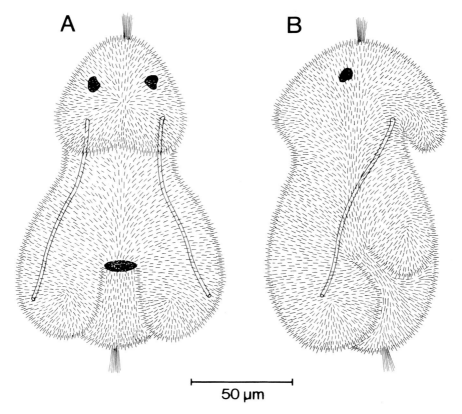

Fig. 1. Larva of *Stylochus mediterraneus* (in A ventral view with the mouth opening, in B lateral view) with one pair of unbranched protonephridia.

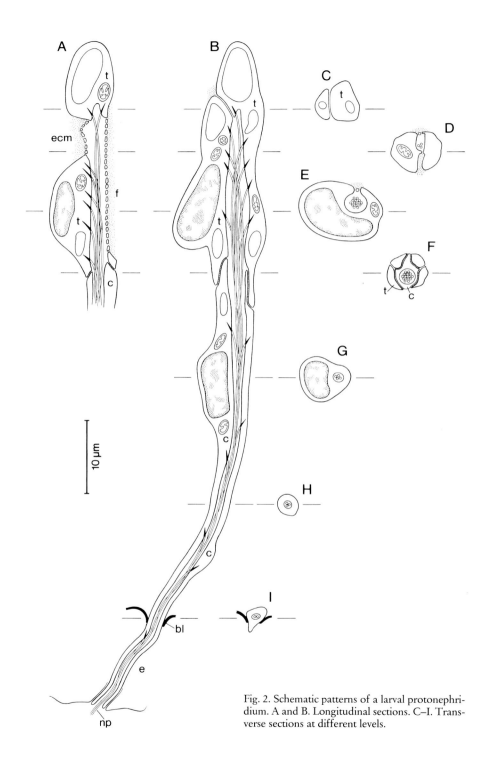

Fig. 2. Schematic patterns of a larval protonephridium. A and B. Longitudinal sections. C–I. Transverse sections at different levels.

cells measures about 5 μm in length, here the posterior parts of the terminal cell envelope the anterior parts of the canal cell (Figs. 4, 7 A, B). Both cells are joined by numerous septate junctions.

1. Terminal cell

The terminal cell is completely splitted 8 μm below its proximal tip (Figs. 2, 3, 4, 5 A, B) into two cytoplasmic parts. The cell gaps are in form of two longitudinal clefts. At one side of the cell, the cleft closes in a distance of about 17 μm beneath the proximal tip just above the nucleus of the terminal cell (Fig. 4). At the opposite side, the other cleft extends to the posterior end of the terminal cell (Figs. 4, 7 A). In this posterior region, several more cell gaps in form of two shorter longitudinal clefts of different lengths can be found (Fig. 4). All the cell gaps become narrow when joining the canal or tubule cell. Here, the narrow gaps display septate junctions between the opposite parts of the terminal cell. (Figs. 4, 7). The central protonephridial lumen communicates with the surrounding intercellular spaces by all these longitudinal cell gaps.

The filter structures of the protonephridium are located in the areas of these longitudinal clefts. Here, more or less irregular microvilli-like cytoplasmic differentiations exist. Several of these microvilli, which extend into the nephridial lumen, may be termed "internal leptotriches". But other microvilli originate from one side or from both sides of the gaps. In the latter case, the microvilli interdigitate forming an irregular system of small slits with a width of about 24 nm (Figs. 2, 5, 6 A, C). The slits between these microvilli or between microvilli and larger cytoplasmic projections (Figs. 5 A, C, 6 C) are covered by extracellular filter diaphragms (Figs. 2, 5, 6 A, C, D). Sometimes these diaphragms also span the longitudinal clefts (Fig. 5 B).

In addition, tongue-like microvilli occur in these filter regions and in the central protonephridial lumen as well. Such a widened microvillus displays a longitudinal row of small vesicles like a string of beads in the central longitudinal axis (Figs. 5 B, 6 D).

The terminal cell envelopes the beginning of the nephridial lumen which has a diameter of about 1.5 μm. About 30 cilia arise almost regularly from the whole adluminal surface of the terminal cell. These cilia measure from 5–14 μm in length and extend distally into the lumen surrounded by the canal cell. Up to 16 cilia can be seen in one cross-section of the terminal cell (Fig. 7 B). Most cilia display two ciliary rootlets, a vertical one with a length of 0.9 μm and a lateral one with a length of 0.3 μm. Several cilia have a vertical rootlet but lack a lateral one. In addition, several centrioles have been found in one of the two distal cytoplasmic projections of the terminal cell (Fig. 7 A).

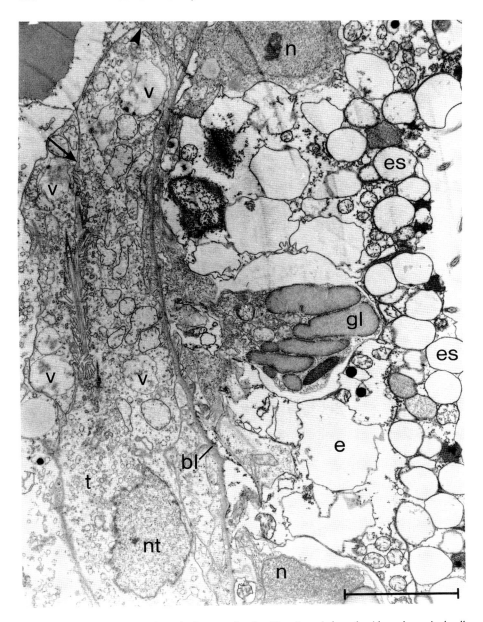

Fig. 3. Longitudinal section through the vacuolated epidermis and the subepidermal terminal cell with the proximal cell tip (arrowhead) and the proximal cell gap (arrow). Scale bar: 5 μm.

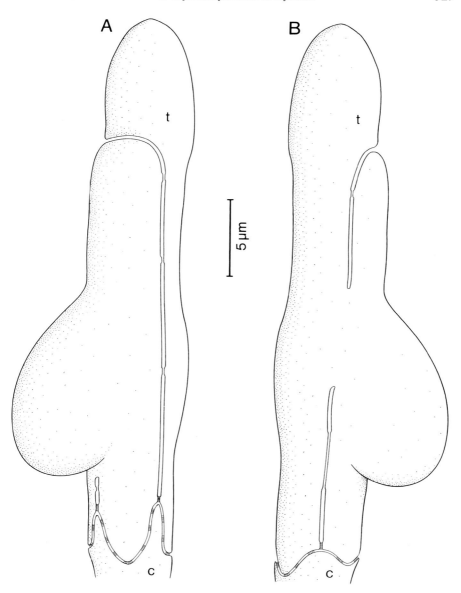

Fig. 4. Schematic patterns of the terminal cell with the different longitudinal cell gaps.

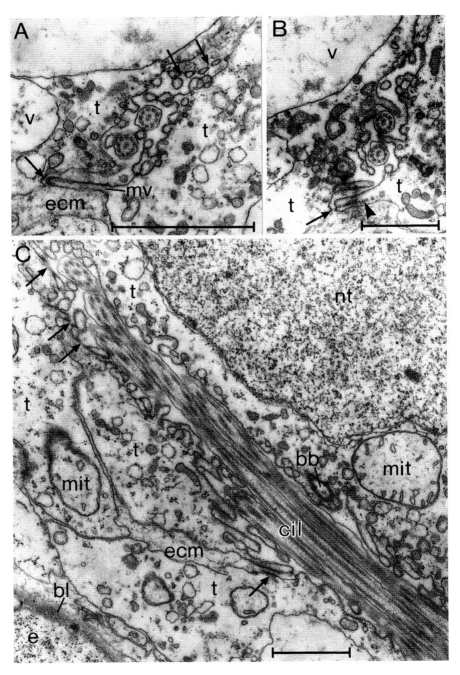

Fig. 5. A and B. Transverse sections of the proximal regions of a terminal cell with filter diaphragms (arrows) and with a longitudinal row of small vesicles (arrowhead in B) in a microvillus-like cytoplasmic extension. In C: longitudinal section of the terminal cell with it nucleus and with filter diaphragms (arrows). Scale bars: 1 μm.

In general, the cytoplasm of the whole terminal cell is relatively electron-lucent (Figs. 3–7). Numerous mitochondria and vesicles, free ribosomes, but only few ER and dictyosomes exist. The nucleus (9 × 5 × 3 µm) takes a position in the more distal region of the cell, here the cell volume is enlarged (Figs. 2–4, 6 A).

Several of the vesicles of the terminal cell reach the size of the nucleus and show flocculent contents (Figs. 3, 5 A, B). Smaller vesicles of this type also exist. Other vesicles of different sizes have an electrondense inner covering of the membrane comparable with the coverings of the membranes adjacent to the nephridial lumen, especially in depressions of the plasma membrane, which represent coated pits (Fig. 6 B). Numerous coated vesicles can be found throughout the cytoplasm, especially near the plasma membrane surrounding the nephridial lumen (Fig. 6 B, but also Figs. 5, 6 C, 7 A).

2. Canal cell and nephroporus

The canal cell or tubule cell is in the form of a close cytoplasmic tube without any cell gap extending along the cell (Figs. 2, 8 B, C). The nucleus lies within the first third of the cell (Figs. 2, 8 B). More distally, the diameter of the canal cell decreases considerably so that the cell gets the appearance of a very narrow tube (Figs. 2, 8 A, C). Sometimes, the distance between the outer and the inner plasma membranes is only 70 nm.

Up to 16 cilia of the terminal cell extend into the lumen surrounded by the proximal parts of the canal cell (Fig. 7 B). The canal cell is more or less uniformly ciliated, but the ciliation is much lower than in the terminal cell. Microvilli were not observed. At the level of the nucleus, only 4 or 5 cilia have been found in cross-sections of the outleading lumen (Fig. 8 B). More distally, the number of cilia varies between 4 to 1 in cross-sections (Fig. 8 C). The plasma membrane of the canal cell surrounding the proximal parts of the lumen displays a conspicuous electron-dense covering (Fig. 7 B).

With its posterior region, the canal cell penetrates the basal lamina and enters the epidermis (Figs. 2, 8 A). Here, in the region of this transition, the basal lamina buldges out and the cytoplasm of the canal cell adjacent to the nephridial lumen disintegrates, because the adluminal plasma membranes are deeply furrowed (Figs. 8 C, 9 A).

The position of the distal intercellular opening of the canal cell, the nephroporus, is marked by a slight depression of the epidermis (Fig. 8 A). Several of the cilia, which arise in this most posterior intraepidermal part of the canal cell, project outside of the nephroporus. These cilia display only one about 3 µm long rootlet. Several centrioles can be found in the vicinity of these rootlets (Fig. 9 A).

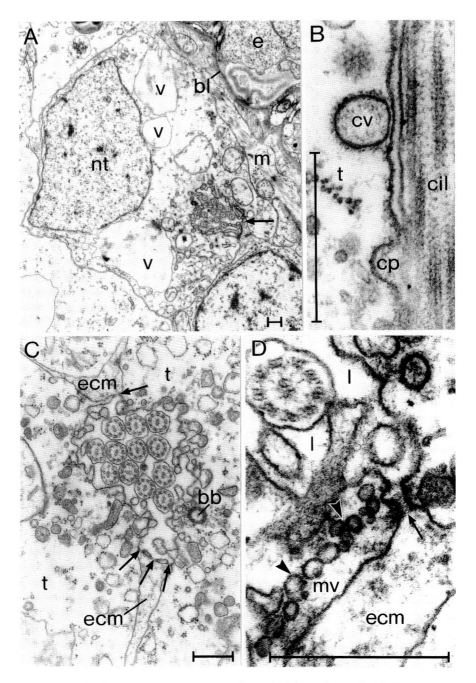

Fig. 6. Terminal cell. In A: transverse section at the level of the nucleus and with filter structures (arrow). In B: longitudinal section with a coated pit and a coated vesicle. In C: transverse section of the distal region with the two longitudinal cell gaps each provided with filter structures (arrows). In D: higher magnification of a microvillus-like cytoplasmic extension with a filter region (arrow) and with a longitudinal row of small vesicles (arrowheads). Scale bars: 0,5 μm.

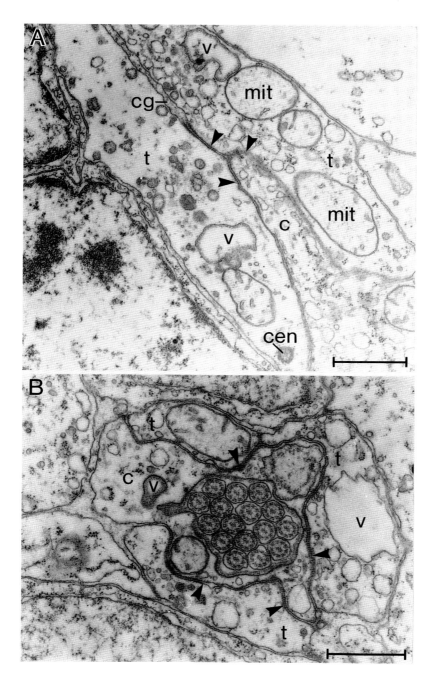

Fig. 7. Longitudinal (in A) and transverse (in B) sections through the transition region between the terminal and the canal cell. Both cells are joined by septate junctions (arrowheads). Scale bars: 1 μm.

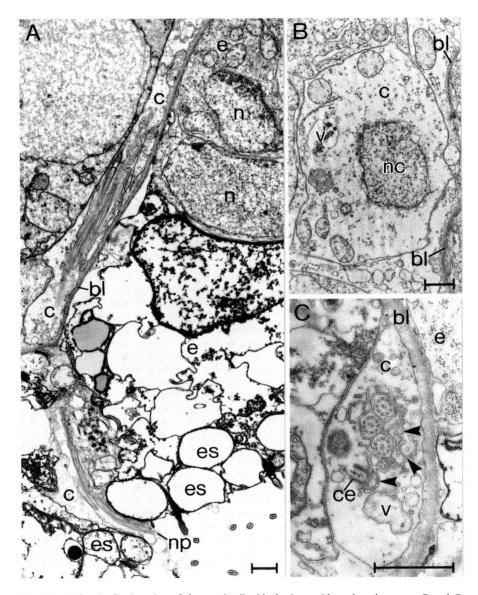

Fig. 8. In A: longitudinal section of the canal cell with the intraepidermal nephroporus. B and C: transverse sections of the canal cell at the level of its nucleus (in B) and at the level just prior to the penetration into the epidermis (in C), here the plasma membranes surrounding the lumen are deeply furrowed (arrowheads). Scale bars: 1 μm.

An additional cell has been found in a larval protonephridium of an increased length and with a more posteriorly located nephroporus (Fig. 9 B, C). At the level of the epidermis, this cell borders one side of the canal cell, both cells are joined by septate junctions (Fig. 9 C). The additional cell, which contains many small vesicles, is quite distinct from the adjacent epidermal cells which display large vacuoles and epitheliosomes (Figs. 3, 8 A, 9 B, C). The additional cell is more similar to the canal cell (Fig. 9 B). But this additional cell lacks any cilia and the apical cytoplasm is disintegrated into many small microvilli-like projections (Fig. 9 B, C).

D. Discussion

The protonephridial system of the Götte's larva investigated here consists of two protonephridia, which do not ramify and which are composed of two cells each, a terminal cell and a canal cell.

Such short protonephridia are unknown from any other plathelminth taxon investigated up to now.

Adult polyclads exhibit complicated protonephridial systems with many branches and with many terminal cell areas and many canal cells (cf. LANG 1884, pp. 164–167, and own EM-findings). RUPPERT (1978, fig. 2 A) found branched protonephridia with canals consisting of several cells in the Götte's larva of another species of the taxon *Stylochus*, most probably, this larva was more developed than the larva of *Stylochus mediterraneus*.

It can be concluded from these available informations, that the short nephridium in the larva of *Styl. mediterraneus* is not a complete one.

This conclusion is corroborated by additional observations:

(1) An additional cell adjacent to the distal parts of the canal cell has been found in a more developed nephridium (Fig. 9 B, C). This additional cell might represent a second canal or tubule cell which becomes integrated into the nephridium or – more probably – develops to a more specialized nephroporus cell (because of the existence of a cell surface elaborated into many microvilli-like differentiations) which is in connection with more than one branch of a more developed system as it is shown by RUPPERT (1978). According to RUPPERT (l. c., fig. 2 B), the larval protonephridium does not become resorbed during metamorphosis.

(2) Whereas the terminal part of a larval protonephridium consists of a single cell (terminal cell), those of adult polyclads are composed of several cells (unpublished observations). In the larva, the filter areas are made by more or less irregular microvilli-like cytoplasmic projections of the terminal cell whereas the

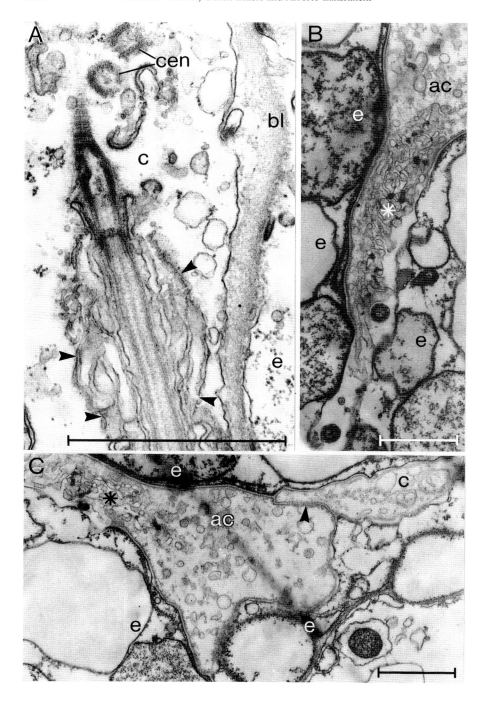

filter in an adult polyclad is much more regular because adjacent terminal cells, which together form a filter region, send off typical microvilli and the microvilli of the adjacent cells interdigitate. The filter microvilli of an adult polyclad are strenghthened by filaments whereas filaments are lacking in the larva. Thus, it can be concluded from yet unpublished EM-observations, that the composed filter area in an adult polyclad originates by the integration of additional terminal cells to an already existing terminal complex which still corresponds well with that of the larva. Hence, our hypothesis is that the larval protonephridia persist during metamorphosis and become more complicated by the addition of more cells (at least in distinct polyclad taxa) during postlarval growth.

(3) Many centrioles have been found in the protonephridial cells of the larva. These observations indicate, that ciliogenesis has not been finished. The larval canal cell is ciliated but lacks any local tuft of cilia (a so-called lateral flame). The existence of such tufts is a common feature for most plathelminth protonephridia (EHLERS 1989) and has been reported for the main vessels of an adult polyclad as well (LANG 1884, p. 166).

But the larval two-cell-nephridium in *Styl. mediterraneus* seems to be able to perform a regular nephridial function, in spite of its early stage of differentiation. There exists a well developed filter region with an extracellular diaphragm, most probably acting as a molecular sieve.

And the terminal cell displays large amounts of typical coated vesicles which originate from coated pits. Most probably, reabsorption takes place in these terminal cells.

The function of the small vesicles arranged to a longitudinal row like a string of beads in many microvilli of the filter remains questionable. But it is obvious, that the most distal vesicle of such a row is quite close to the extracellular diaphragm.

Historically, there is the notice of LANG (1884, p. 355 and table 36, figs. 1–8), that the two protonephridia in a polyclad embryo might develop from two caudal invaginations of ectodermal cells. According to LANG, these cells form club-like differentiations which again penetrate into the epidermis in a later developmental stage. The description of LANG compares well with the findings presented in this paper: (a) The larval nephridium is located just beneath the subepidermal basal lamina and the subepidermal musculature. (b) The distally

◀ Fig. 9. In A: higher magnification of a longitudinal section through the canal cell penetrating the basal lamina. Arrowheads point to deeply forrowed membranes surrounding the nephridial lumen. In B and C: transverse sections of the intraepidermal parts of an additional cell with its apical microvilli (asterisks); additional cell and canal cell are joined by septate junctions (arrowhead in C). Scale bars: 1 μm.

orientated buldge of the basal lamina in the vicinity of the canal cell can be explained by a secondary penetration of this cell into the epidermis.

In summary, there is strong evidence that the larval nephridium in *Stylochus mediterraneus* is not a completely differentiated one. Thus, it does not make sense to compare this larval nephridium with the known nephridia in adult specimens of other plathelminth taxa in order to discuss phylogenetic aspects.

Nevertheless, these statements can be given:

(1) One pair of protonephridia and two nephropori exist in the larva. Both features are plesiomorphic characteristics, inherited from the stem species of the Plathelminthes or the Bilateria, respectively (Ax 1984, 1987; EHLERS 1985).

(2) The protonephridium of the adult stem species of the Bilateria is hypothesized to consist of three cells (Ax 1984, 1987): a terminal cell, a canal cell and a nephroporus (or distal canal) cell. This hypothesis is not in conflict with the new findings on the larval nephridium in a polyclad: the two-cell-nephridium is an early, not yet completely differentiated organ, but it is working, because the Götte's larva represents an early stage of development, an embryo, which has become an active pelagic individual (EHLERS 1985).

(3) This explanation also holds true for the fact that the two-cell-nephridium is not yet branched. Branched nephridia with more than one terminal cell can be hypothesized for the adult stem species of the Plathelminthes (EHLERS 1985).

(4) The situation, that the filter in the larval nephridium is formed by one cell, a terminal cell, represents a plesiomorphic characteristic. Such a situation is hypothesized for the stem species of the Plathelminthes (EHLERS 1985).

(5) The canal cell in the larva is in the form of a closed tube. Such a canal cell also exists in the larva figured by RUPPERT (1978, fig. 5 D). Canal cells with a longitudinal cell gap exist in an adult polyclad (unpublished observations). Both situations are known from other plathelminth taxa and can be found in one specimen (EHLERS 1989). The organization of the canal system can apparently vary in several taxa of the Rhabditophora (Macrostomida, Polycladida, Prolecithophora), whereas it seems to be more stable in other taxa (Proseriata, free-living and parasitic Rhabdocoela). Comparable changes in the organization of the canal cells (with or without a cell gap) during development are also known from the Gastrotricha (NEUHAUS 1987, p. 421).

Acknowledgements

The authors should like to thank Mr. B. Baumgart for the realization of the drawings. This research was supported by the Akademie der Wissenschaften und der Literatur, Mainz, and by Consiglio Nazionale delle Ricerche, Gruppo Nazionale Scienze Neurologiche.

Zusammenfassung

Die Götte'sche Larve von *Stylochus mediterraneus* Galleni, 1976 besitzt ein Paar unverzweigter Protonephridien, bestehend aus jeweils 2 Zellen: eine Terminal- und eine Kanalzelle. Die Protonephridien erstrecken sich in der Körperlängsachse unterhalb der Basallamina und münden caudal durch den von der Kanalzelle gebildeten Nephroporus getrennt aus.

Die Terminalzelle ist proximal gespalten. In der Mitte ist dieser Spalt zum Nephridiallumen hin erweitert. Der Längsspalt schließt sich einseitig proximal der Kernregion und bleibt auf der gegenüberliegenden Seite bis zur Kanalzelle erhalten. Bis zu 30 Cilien und unregelmäßig geformte Mikrovilli entspringen von der gesamten luminalen Zellmembran.

Der eigentliche Filter befindet sich mehr oder weniger tief in den Spalten zwischen einander gegenüberliegenden Zellmembranen, die hier zumeist unregelmäßig geformten Mikrovilli angehören. Filterdiaphragmen sind zwischen diesen teilweise ineinander ragenden Mikrovilli oder auch zwischen den Zellwänden selbst ausgebildet.

Die geschlossen rohrförmige Kanalzelle, die ebenfalls Cilien aufweist, durchbricht distal die Basallamina zur Epidermis. In diesem Abschnitt ist die luminale Zellmembran zum Cytoplasma hin gefaltet.

Eine zusätzliche Zelle wurde in einem weiter entwickelten larvalen Nephridium gefunden. Diese Zelle, die im Bereich des Nephroporus mit der Kanalzelle eng verbunden ist, mag eine künftige Nephroporuszelle oder eine weitere Kanalzelle darstellen. Sehr wahrscheinlich entwickelt sich das larvale 2-Zell-Nephridium zum vielzelligen und verzweigten Protonephridium adulter Polycladen, indem weitere Zellen in das bereits in der Larve existierende Nephridium integriert werden.

Abbreviations

ac	additional cell	cv	coated vesicle	mv	microvillus (microvilli) of the filter
bb	basal body	e	epidermis		
bl	basal lamina	ecm	extracellular matrix	n	nucleus
c	canal cell	es	epitheliosome(s)	nc	nucleus of canal cell
cen	centriole	f	filter	np	nephroporus
cg	cell gap(s)	gl	gland cell	nt	nucleus of terminal cell
cil	cilium (cilia)	m	muscle cell	t	terminal cell
cp	coated pit	mit	mitochondrion	v	vacuole(s)

References

Ax, P. (1984): Das phylogenetische System. Systematisierung der lebenden Natur aufgrund ihrer Phylogenese. G. Fischer, Stuttgart, New York, 349 p.
- (1987): The phylogenetic system. The systematization of organisms on the basis of their phylogenesis. John Wiley & Sons, Chichester, 340 p.

Bresslau, E. (1928–1933): Turbellaria. In: Handbuch der Zoologie, gegr. von W. Kükenthal, Vol II/1 (Ed.: T. Krumbach). Walter de Gruyter, Berlin, 52–304.

Ehlers, U. (1985): Das phylogenetische System der Plathelminthes. G. Fischer, Stuttgart, New York, 317 p.
- (1989): The protonephridium of *Archimonotresis limophila* Meixner (Plathelminthes, Prolecithophora). Microfauna Marina **5**, 261–275.

Hyman, L. H. (1951): The Invertebrates. Platyhelminthes and Rhynchocoela. McGraw-Hill, New York, 550 p.

Lacalli, T. C. (1983): The brain and central nervous system of Müller's larva. Ca. J. Zool. **61**, 39–51.

Lanfranchi, A., C. Bedini & E. Ferrero (1981): The ultrastructure of the eyes in larval and adult polyclads (Turbellaria). Hydrobiologia **84**, 267–275.

Lang, A. (1884): Die Polycladen (Seeplanarien) des Golfes von Neapel und der angrenzenden Meeresabschnitte. Fauna und Flora des Golfes von Neapel 11, W. Engelmann, Leipzig, 688 p. and 39 Tables.

Neuhaus, B. (1987): Ultrastructure of the protonephridia in *Dactylopodola baltica* and *Mesodasys laticaudatus* (Macrodasyida). Implications for the ground pattern of the Gastrotricha. Microfauna Marina **3**, 419–438.

Prudhoe, S. (1985): A monograph on polyclad Turbellaria. Oxford University Press, Oxford, 259 p.

Ruppert, E. E. (1978): A review of metamorphosis of turbellarian larvae. In: Settlement and metamorphosis of marine invertebrate larvae (Eds.: F.-S. Chia & M. E. Rice), Elsevier, New York, 65–81

Dipl.-Biol. Cornelia Wenzel and Prof. Dr. Ulrich Ehlers
II. Zoologisches Institut und Museum der Universität Göttingen
Berliner Straße 28, D-3400 Göttingen

Prof. Dr. Alberto Lanfranchi
Dipartimento di Scienze del Comportamento Animale e dell'Uomo, Universita' di Pisa,
Via A. Volta 6, I-56100 Pisa.

Short communication

Plathelminthes from brackish water of Northern Japan: No identical species with the corresponding boreal community

Peter Ax

The species composition of brackish-water Plathelminthes in Northern Europe, Canada (Atlantic Coast) and Alaska (Pacific Coast, Bering Street) shows remarkable similarities resulting from a high number of identical species. The most parsimonious causal explanation of this phenomenon is the hypothesis that a species-rich community of brackish-water plathelminths has a northern circumpolar distribution (Ax & ARMONIES 1987, 1990; Ax 1991).

On the other hand the southern limits of this boreal community were unknown till now. We attacked the problem in Northern Japan. Working at the University of Hirosaki in autumn 1990 I studied about 20 species from brackish water biotopes of Honshu, Aomori Prefecture. Samples were collected at the Japanese Sea (Jusan Lake), at Mutsu Bay (Kominato River, Noheji River) and alongside the Pacific Ocean (Ogawara Lake, Takase River with a brackish water lagoon, Takahoko Pond).

Most of the analysed species belong to well known taxa like *Macrostomum* (Macrostomida), *Minona, Duplominona, Archiloa* (Proseriata), *Multipeniata* (Prolecithophora), *Promesostoma, Phonorhynchoides* and *Pogaina* (Rhabdocoela).

In no case however structural conformities exist with any member of the boreal community of brackish-water plathelminths allowing hypotheses about species identities. The species structure of the two communities being under discussion is totally different. That means that the southern limit of the boreal community is crossed in this part of the world.

Cordially thanks to Prof. Dr. W. Teshirogi, Prof. Dr. S. Ishida, Mrs. K. Ohkawa, Hirosaki University and Dr. S. Takeda, Dr. R. Kuraishi, Marine Biological Station Asamushi of Tohoku University for their support of our research work.

Literature

Ax, P. (1991): Northern circumpolar distribution of brackish-water plathelminths. Hydrobiologia **227**, 365–368.

Ax, P. & W. Armonies (1987): Amphiatlantic identities in the composition of the boreal brackish water community of Plathelminthes. A comparison between the Canadian and European Atlantic Coast. Microfauna Marina **3**, 7–80.

Ax, P. & W. Armonies (1990): Brackish water Plathelminthes from Alaska as evidence for the existence of a boreal brackish water community with circumpolar distribution. Microfauna Marina **6**, 7–109.

Professor Dr. Peter Ax
II. Zoologisches Institut und Museum der Universität Göttingen
Berliner Straße 28, D-3400 Göttingen

Microfauna Marina

Veröffentlichung der Akademie der Wissenschaften und der Literatur, Mainz
Herausgegeben von Prof. Dr. P. Ax, Göttingen
Mit Beiträgen zahlreicher Fachautoren in deutscher und englischer Sprache

Volume/Band 1
1984. X, 277 S., geb. DM 89,–

Aus dem Inhalt: Besiedlungsstruktur freilebender Plathelminthen im Sandwatt, in einer instabilen Sandbank und in einem stabilen Strandhaken der Insel Sylt. Beschreibungen von Copepoden-, Nematoden-, Anneliden- und Plathelminthenarten aus dem Sand der Galápagos-Inseln.

Volume/Band 2
1985. 410 S., geb. DM 98,–

Aus dem Inhalt: Ultrastruktur-Befunde erstrecken sich auf Statocyste und Hautdrüsen freilebender Plathelminthes, auf den Pharynxapparat interstitieller Polychaeta und das Protonephridialsystem der Nemertini. Ökologische Aspekte behandeln Arbeiten über die Meiofauna der Sulfidschicht des Sandwatts, über die Plathelminthen-Assoziationen des Mudwatts und der sublitoralen Sande von Sylt. Systematische Studien liefern Beschreibungen zahlreicher neuer Arten aus den Taxa Copepoda, Nematoda, Oligochaeta und Plathelminthes.

Volume/Band 3
1987. 438 S., geb. DM 98,–

Aus dem Inhalt: Die Ökologie freilebender Plathelminthen in supralitoralen Salzwiesen der Nordsee und im Grenzraum Watt-Salzwiese lenitischer Gezeitenküsten. Beschreibungen neuer Plathelminthen aus dem Brackwasser der Insel Sylt, von Bermuda und der Küste von South Carolina. Ultrastrukturelle Untersuchungen über Plathelminthen (z. B. des Photorezeptors von Macrostomum spirale) und Gastrotrichen.

Preisänderungen vorbehalten

Volume/Band 4 · The Ultrastructure of Polychaeta
1988. 494 pp., hard cover DM 118,–

33 experts review the entire ultrastructural literature on Polychaeta, and present results of their current studies. These data are examined to see how they confirm or change traditional concepts about the polychaetes' morphology, embryology, physiology, and even ecology. One major aim of the book is the use of ultrastructural information to resolve questions about phylogenetic relationships not only between Polychaetes, but between Polychaeta, other Annelida, and other invertebrate taxa.

Volume/Band 5
1989. 329 pp., hard cover DM 94,–

Aus dem Inhalt: Kalyptorhynchia (Plathelminthes) from Sublittoral Coastal Areas near the Island of Sylt (North Sea). I. Schizorhynchia · Coelogynopora visurgis nov. spec. (Proseriata) und andere freilebende Plathelminthes mariner Herkunft aus Ufersanden der Weser · Interstitielle Fauna von Galapagos. XXXVI. Tetragonicipitidae (Harpacticoida) · ... XXXVII. Metidae (Harpacticoida) · ... XXXVIII. Haloplanella Luther und Pratoplana Ax (Typhloplanoida, Plathelminthes) ...

Volume/Band 6
1990. 272 pp., num. figs. and tabs., hard cover DM 89,–

Aus dem Inhalt: Brackish water Plathelminthes from Alaska as evidence for the existence of a boreal brackish water community with circumpolar distribution · On the morphology and amphiatlantic distribution of Jensenia angulata (Jensen, 1878) (Dalyelliidae, Plathelminthes) · Feinstrukturelle Untersuchungen an Vitellarien und Germarien von Coelogynopora gynocotyla Steinböck, 1924 (Plathelminthes, Proseriata) ...

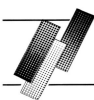

BIONA report

Herausgegeben von Prof. Dr. Werner Nachtigall, Saarbrücken
Schriftenreihe der Akademie der Wissenschaften und der Literatur, Mainz

Report 1	**Physiologie und Biophysik des Insektenfluges** Atmung, Stoffwechsel, Flügelbewegung 1983. X, 135 Doppelseiten, zahlr. Abb., kt. DM 38,–
Report 2	**Physiologie und Biophysik des Insektenfluges** Neuro-, Sinnes- und Muskelphysiologie 1983. X, 137 Doppelseiten, zahlr. Abb., kt. DM 38,–
Report 3	**Bird flight/Vogelflug** 1985. XVIII, 509 S., 8 Taf., zahlr. Abb. u. Tab., kt. DM 48,–
Report 4	**Temperature Relations in Animals and Man** Edited by Prof. Dr. H. Laudien, Kiel 1986. X, 234 pp., soft cover DM 34,–
Report 5	**Bat flight/Fledermausflug** 1986. XII, 235 S., zahlr. Abb. u. Tab., kt. DM 38,–
Report 6	**The Flying Honeybee (Aspects of Energetics)/Die fliegende Honigbiene (Aspekte der Energetik)** Herausgegeben von Prof. Dr. W. Nachtigall, Saarbrücken 1988. X, 151 S., zahlr. Abb. und Tab., kt. DM 28,–
Report 7	**3-D SEM-Atlas of Insect Morphology** Vol. 1: Heteroptera By Dr. H. G. Kallenborn, Dr. A. Wisser, and Prof. Dr. W. Nachtigall, Saarbrücken 1990. XII, 164 pp., 38 figs., 134 photographs (incl. a pair of prismatic spectacles), soft cover DM 54,–
Report 8	**Technische Biologie und Bionik 1** 1. Bionik-Kongreß, Wiesbaden 1992 Herausgegeben von Prof. Dr. W. Nachtigall, Saarbrücken 1992. X, 168 S., zahlr. Abb. u. Tab., kt. DM 44,–

Preisänderungen vorbehalten

Bestellkarte

Ich bestelle aus dem Gustav Fischer Verlag, Stuttgart, über die Buchhandlung:

....

Microfauna Marina

30460	.... Ex.	—, Band 1, DM 89,—
30490	.... Ex.	—, Band 2, DM 98,—
30558	.... Ex.	—, Band 3, DM 98,—
30581	.... Ex.	—, Vol. 4, Polychaeta, DM 118,—
30608	.... Ex.	—, Band 5, DM 94,—
30663	.... Ex.	—, Band 6, DM 89,—

BIONA report

20300	.... Ex.	Report 1, DM 38,—
20001	.... Ex.	Report 2, DM 38,—
20330	.... Ex.	Report 3, DM 48,—
20357	.... Ex.	Report 4, DM 34,—
20372	.... Ex.	Report 5, DM 38,—
20434	.... Ex.	Report 6, DM 28,—
20467	.... Ex.	Report 7, Vol. 1, DM 54,—
30696	.... Ex.	Report 8, DM 44,—

Preisänderungen vorbehalten

Datum: .... Unterschrift: ....

Zur Information über Neuerscheinungen und Neuauflagen des GUSTAV FISCHER VERLAGS auf Ihrem Fachgebiet schicken wir Ihnen auf Wunsch laufend kostenlos Informationen zu. Interessengebiete bitte ankreuzen und Karte ausgefüllt zurückschicken.

Medizin
- ☐ Anatomie, Embryologie
- ☐ Pathologie
- ☐ Physiologie
- ☐ Med. Mikrobiologie, Hygiene
- ☐ Pharmakologie, Toxikologie
- ☐ Pharmazie
- ☐ Labormedizin
- ☐ Innere Medizin, Allgemeinmedizin
- ☐ Anästhesie, Intensivmedizin
- ☐ Chirurgie, Orhopädie, Urologie, Röntgenologie, Sonographie, NMR, diagnostische Nuklearmedizin
- ☐ Gynäkologie, Geburtshilfe, Perinatologie
- ☐ Pädiatrie, Perinatologie
- ☐ Ophthalmologie
- ☐ Oto-Rhino-Laryngologie
- ☐ Dermatologie, Venerologie
- ☐ Zahnheilkunde
- ☐ Neurologie
- ☐ Psychiatrie, Psychotherapie
- ☐ Psychologie
- ☐ Musiktherapie
- ☐ Medizinalfachberufe, Physikal. Medizin, Krankenpflege, Krankengymnastik, Massagen, MTA
- ☐ Rechtsmedizin, Arbeits- und Sozialmedizin, Begutachtung
- ☐ Gesch. der Medizin und Naturwissenschaften

Biologie
- ☐ Veterinärmedizin
- ☐ Umwelthygiene
- ☐ Botanik (incl. Ökologie, Allg. Biologie, Biogeographie)
- ☐ Zoologie (incl. Ökologie, Allg. Biologie, Mikrobiologie, Biogeographie)
- ☐ Anthropologie, Ethnologie, Evolution, Paläontologie
- ☐ **Statistik, Biometrie, Datenverarbeitung**
- ☐ **Wirtschafts- und Sozialwissenschaften**

Absender:
(Studenten bitte Heimatanschrift angeben)

....

....

....

Ich bitte um kostenlose Zusendung von
☐ Teilverzeichnis Biologie

Falls keine Buchhandlung bekannt, bitte einsenden an:

Gustav Fischer Verlag
Postfach 72 01 43
W-7000 Stuttgart 70

Microfauna Marina, Vol. 7,
VI. 92. 0,55. nn. Printed in Germany

Bitte ausreichend frankieren

Werbeantwort/Postkarte

An die Buchhandlung

Absender:

....

....

....

Kostenlos
☐ Verzeichnis Biologie

Microfauna Marina, Vol. 7,
VI. 92. 0,55. nn. Printed in Germany

Bitte ausreichend frankieren

Werbeantwort/Postkarte

Gustav Fischer Verlag
Postfach 72 01 43

D-7000 Stuttgart 70